D0097905

LANGUAGES OF THE ANIMAL WORLD

Also by J. H. Prince

Animals in the Night: Senses in Action After Dark
The Universal Urge: Courtship and Mating Among Animals
Comparative Anatomy of the Eye
The Rabbit in Eye Research
Weather and the Animal World

LANGUAGES OF THE ANIMAL WORLD

by

J. H. Prince

THOMAS NELSON INC., PUBLISHERS
Nashville / New York

Published in Nashville, Tennessee, by Thomas Nelson Inc., Publishers, and simultaneously in Don Mills, Ontario, by Thomas Nelson & Sons (Canada) Limited. Manufactured in the United States of America.

First edition

Library of Congress Cataloging in Publication Data

Prince, Jack Harvey.
Languages of the animal world.

Includes index.
Bibliography: p.
1. Animal communication. I. Title.
QL776.P74 571.5′9 75–19464
ISBN 0–8407–6470–7

CONTENTS

LANGUAGES OF THE ANIMAL WORLD

INTRODUCTION

Almost all living creatures communicate with their own species either through sounds, through movement, through body posture, or through odor, and man has inherited the ability to use all of these methods too. But only in this century—mostly in the last thirty years—has it been realized to what extent a study of the ways in which animals communicate with each other can help us to understand our own strange ways.

There is perhaps a tendency to think of human communication only in terms of conversation, but we also use hand gestures, eyebrow movements, and in fact every kind of muscular movement to express our thoughts and intentions. We don't use our body odors for intentional message-carrying as many lower animals do, but we *do* use artificial odors to give information. Using perfume is an easy way for a woman to say without words that she wants to be attractive.

Nor is human language confined to words alone. Like countless animals we convey meaning with grunts, gasps, and rapid drawing in of breath. These can be accompanied by blushing, muscle twitching, head nodding and so on, none of which we may be aware of, but they still convey our mood or meaning. There is, however, one very great language difference between man and the other animals. Only man uses what we call *words*, which can be strung together into sentences that will tell a story of the past, the present, or the future, or that will build up an imaginary story or present an abstract idea.

Only man has devised a way of recording his language permanently with written symbols. Because he has been able to do this, his messages can be sent over great distances without his being present or even alive. His words can also be stored in print for an indefinite time. No other animal can send a message a great distance without being present, and no other animal can leave any kind of permanent record. They must be within sight or hearing distance of each other to send and receive messages. Even when an animal marks a place with its odor to prove ownership, the odor lasts only briefly.

There are both advantages and disadvantages to each of the three forms of communication used by the animal world. The use of sound for sending a message permits a rapid exchange of signals between two individuals or groups and also permits the correcting of wrong signals, so sound is valuable to a species that moves around rapidly, especially if it inhabits territory where vision is restricted. At the same time, sound messages need not interrupt other activities such as eating, whereas vision demands full attention.

Sounds can be given very different meanings by slight variations. For instance, our words "vivid" and "livid," which might sound identical to our dogs, mean very different things to us. Another example might be "cuff-links" and "golf-links" or "cheap" and "cheat." Different sequences of similar sounds have different meanings for animals just as they have for us. For example, we understand a great difference between "John hits Bill" and "Bill hits John," even though exactly the same words are used. There are also times when a number of sounds strung together will have meaning whereas a single sound will not. The same thing applies to body positions. A series of positions followed in a certain order will mean something quite different from the same positions followed in another order. The significance of a sound or a position can also be changed by giving it emphasis.

If we try to think of all the things we are likely to say throughout a day, we soon see that many of them are messages—and sometimes important messages, such as "look out!" "come here," "stop it," "I'm hungry," "I've found food," or "I feel sick." We might say "look out" in a normal voice or we might shout it to emphasize urgency in a more dangerous situation. Since everything we say can be varied in tone, we say many different things in different ways at different times.

Animals need to do this too, but we seldom understand more than a tiny fraction of their language. This is because so much of it sounds the same to us—like the cry of a human baby, which sounds the same to most of us whether it is crying because it is hungry or because it is just wet.

Animals unable to express themselves with a voice express themselves in other ways. Whatever way it may be, it is sufficiently effective for the species to survive and prosper.

When we look carefully at the subject of language, we see that man in fact uses several languages, all of which are shared in some way with the rest of the animal world. They are the result of voice, gesture, body attitude, facial expression, odor, and in some situations the complete absense of sound or gesture. Each of these carries its own special significance.

It has been suggested that no animal intentionally communicates with another. The animal merely expresses its feelings, and others see, hear, or smell this. It is even possible that an animal may not be aware of the fact that it is communicating by giving a signal or making sounds, because it will do this even when there is no other member of the species around to hear or see it. (A bird will give a warning call, for instance, or a deer will flash its tail.) This is different from our own idea of communication, which is usually *intended* to influence another's thoughts or actions, or to give them an idea of how we ourselves are feeling or thinking.

The many forms of language or expression in lower orders are important to their survival and in fact have evolved for that very purpose. We can see this clearly in signals that are related to a tail or ears, because these signals are genetically controlled. The white underside of a rabbit's tail can spell survival instead of doom for another rabbit that sees it flash a danger signal. This applies to many other animals because tail-raising at various levels is common in all mammals with tails.

Any animal deficient in a part of its anatomy that is essential for signaling or any animal that is unable to make the correct sounds will be less successful in finding a mate, so it is obvious that accurate communication is essential to the survival of a whole species. Because we are aware of so little that goes on around us in the animal world, many of us only recognize that a dog barks, a horse stamps its foot, a cat hisses, a bird sings, and so on.

There are languages that include sounds beyond our range of hearing, as well as sounds that we can hear. There are languages of signals made by body movements and postures, through touch, with light, with dance patterns, or with odors that are borne on the wind or that are deliberately placed on objects or other animals.

Man is able to learn and interpret some aspects of these languages, just as he can learn the languages of other human groups. But animals are unable to understand the language of any other species; they can understand only their own dialect, as though it were a computer code. Nevertheless, many animals understand the warning or alarm signals of other species because of the tone used.

Most mammals and birds seem able to sense alarm in others, and this often enables different kinds of animals to mingle together for greater safety. Such an association is found in animals such as the baboon [90] and the impala,[4] which find mutual protection by staying close together, because the warning signals of each are understood by the other.

There are many situations in which communication must be silent if the unwelcome attention of predators is to be avoided. Warnings to the rest of a group or to the family are achieved in open country or when individuals are close to each other by movement and position of the tail, by body attitudes, or by movement of the ears. Though silent, this is still language.

Warnings to enemies need not be silent, but frequently they are. When an enemy has approached within what is called an animal's "safe distance," the silent warning is by facial expression, the attitude of horns, tail, or the whole body, and it precedes attack. One of the elements of survival is naturally the correct interpretation of warning signals such as the spreading of an animal's nostrils or the change of eye position, both of which precede certain animals' attacks.

Gestures, calls of affection, scolding, fear, anger, expressions of indifference, contempt, pain, and unhappiness are all seen in many animals. Appeals for help are even responded to just as they are by human beings. Some animal language patterns are in fact exceedingly complex. Just as isolated human tribes or nations have developed different languages, so groups of animals of the same species that have been separated from each other for thousands of years by deserts or mountains have also developed somewhat different languages or dialects.

One of the basic functions of many animal calls and signals is to help individuals locate or avoid each other. In some situations these signs may also indicate such things as "up" or "down," so signals must be precise for quick understanding. Since this precision is especially necessary for alarm signals among hunted animals, we find that many animal languages are therefore quite sophisticated.

As well as searching into the meanings of the different patterns of animal communication, we must understand how sounds, light, and odors are received and made use of and also how they are transmitted by animals in many different ways. One part of this book will deal with these concepts. The other parts will explain just how animals use these abilities, and what the various signals may mean. Species names are listed in a glossary at the end of the book, according to numbers beside the common names of animals as the names appear in the text.

PART I
WAYS IN WHICH ANIMALS COMMUNICATE

1. How Animals Hear

Animals would never have developed hearing if there had not been any sound. But even before the appearance on earth of animals there were always sounds. First there were natural sounds of things like wind, water, rustling foliage, and then as animals evolved, there were the sounds of creatures moving through the water, through aquatic plants, and then over sand and through grass or other foliage.

Hearing probably first evolved as animals made use of these natural sounds. Then when the beginnings of true hearing became possible, the advantages of this sense encouraged animals to develop special ways of making sounds, because this improved their chances of survival and of finding mates. The very beginnings of true speech were at that time a long way in the future, but speech had become both possible and inevitable.

We cannot be sure if true hearing was first developed for detecting approaching enemies, for identifying members of the same species, purely for courtship, or for another purpose, but it seems logical to assume that it was for the detection of approaching enemies. Apparatus for receiving sounds can obviously exist in an animal without that animal being able to send sound signals itself, but the animal would never use sound either through a voice or through tapping a foot unless the animal also had some kind of receiving apparatus, such as an ear, with which to receive sound.

"Hearing" can be described simply as the reception of

different vibrations by special organs that change the vibrations into electrical energy and send the energy to the brain, where it is "understood," that is, the meaning is intelligently understood by man and some of the higher vertebrates, but in most of the lower orders the sound is merely responded to by actions away from or toward it.

We cannot be sure that a particular sound will produce the same effect in the hearing organs in the leg of a grasshopper as it will in the ear of a man, because the grasshopper's "ears" and brain are vastly different from man's. Nor do all kinds of animals have the same range of hearing. The cat, for instance, is sensitive to higher-pitched sounds than humans are.

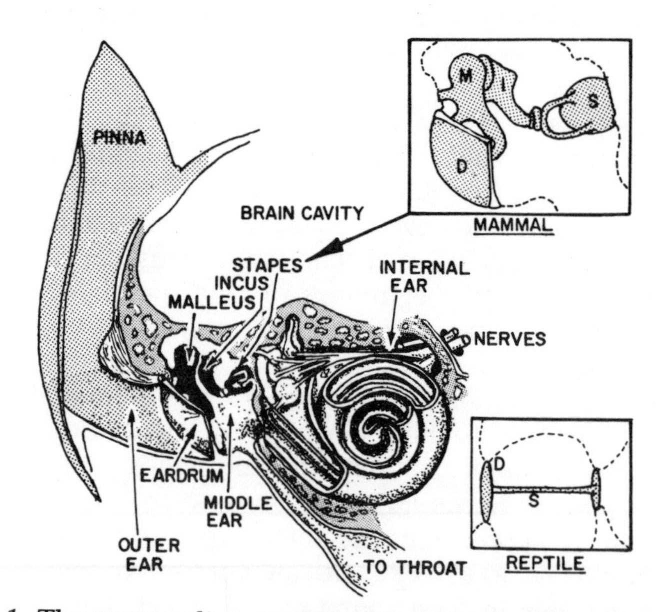

Figure 1. The mammalian ear has three chambers, an outer ear and *pinna* for collecting sound and directing it onto an *eardrum* that vibrates three small bones (black) in a middle ear chamber, which in turn transmit the vibrations to an inner ear where they stimulate nerves to the brain. The reptilian ear (inset) is a much simpler form with a single *stapes* bone (S).

Vertebrate Ears. Not all ears are the same. Those of mammals and many other land vertebrates are located within the skull, with openings to the outside. Those of fish are more primitive, sealed entirely within the skull and often connected to a drumlike gas bladder in the body cavity. In insects organs of hearing may be located in the legs, the abdomen, the thorax, or the antennae.

The most advanced vertebrate ears are divided into three chambers—outer, middle, and inner. Sound vibrations are collected by a *pinna*, the outer shell of the ear, which can be seen and which acts like a dish antenna to direct the sound vibrations into the outer ear canal (Figure 1), at the inner end of which is a taut eardrum called a *tympanum.* The pressure created on the eardrum by the vibrations causes the drum to vibrate too, but this drum vibration is so slight that it can be measured only in millionths of an inch. Nevertheless, it is enough to agitate the *ossicles* (small bones) in the middle ear chamber behind the drum.

There are three such bones in the middle ear chamber of mammals. They are known as the *malleus* (hammer), which is in contact with the back surface of the eardrum; the *incus* (anvil), which joins the malleus to the third bone; and the *stapes* (stirrup), which is shaped like a rider's stirrup. The "footplate" of the stapes covers a little oval window between the middle and inner ear chambers, so the bones conduct the eardrum vibrations right to the inner ear. This oval window has an area which in human beings is only 3.5 percent of the eardrum, so sound is amplified more then twenty-seven times. The size of this window varies in other animals according to their needs.

There is a small *stapes muscle* that prevents the eardrum from vibrating beyond a certain limit, and this muscle has developed so that very loud noises will not damage the delicate ear structure. In the middle ear chamber is an opening that leads into a tube—the *eustachian tube*—connecting it with the throat. This tube permits air to reach the middle ear, thus equalizing the air pressure on both sides of the eardrum.

The inner ear is a very complex arrangement of fluid chambers and canals that encloses the sensitive endings or "receivers" of the auditory (hearing) nerves to the brain. Part of this complex system also controls the balancing power of the body. When sound vibrations reach the nerve receptors, they produce electrical impulses, and the tiny currents thus created pass to the brain centers, where they are interpreted as sound. In exactly the same way, the receptors in the retina of the eye produce electrical impulses in response to light vibrations.

The Task of the Brain. Two kinds of pathways are taken by nerves from the ears to the brain. One of these probably helps the ear to function with the eye—something that is very important when trying to locate an enemy quickly. Somewhere in the brain these systems are also connected with the muscles that turn the head toward the direction of sound. This system of nerves and its brain center are different in nocturnal animals from animals active only during the day, because day animals use their eyes more than nocturnal ones.

The hearing centers in the brain must be able to do many things, not all of which are related solely to *receiving* sounds. The hearing centers must analyze mixtures of sounds and must learn to ignore background noises and pick out only the sounds that are important at the moment. This is like listening to conversation in a noisy train. We try to shut out everything except what is being said. These centers must identify direction by rapid measurement of the difference in the volume of sound reaching the two ears. Adjustments must be made for loudness; and memory and experience must also be used to give a correct meaning to what is heard. No computer could do better.

Because the pinna gathers sound like the mouth of a trumpet and directs the sound into the ear passage to the eardrum, the pinna is almost always large in animals active at night. The larger the pinna is, the more sound it gathers.

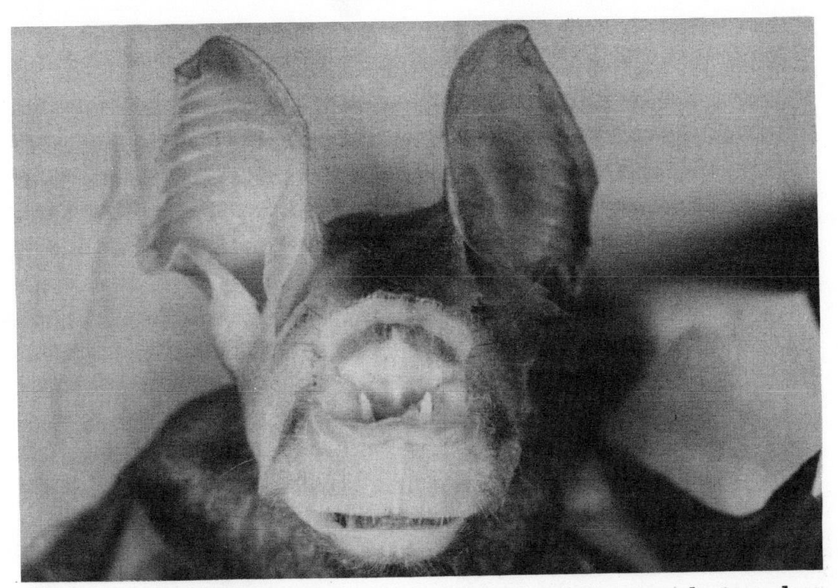

Figure 2. The giant pinnas of this bat's [51] ears are designed to pick up the faintest sounds. Its nose is constructed to direct sounds ahead of it and to amplify them.

Its shape is important too, because the effects of certain sounds are changed by being directed into the ear canal at different angles. Experiments on animals in which the pinnas of the ears have been removed surgically and the brain responses to sounds have been measured electronically have shown that the level of hearing is greatly reduced and that judgment of direction becomes much less accurate.

One of the brain's most valuable qualities is its ability to judge the distance and direction of a sound. Unless a sound is immediately overhead or in perfect line with the direction of the body, the sound is heard sooner and more loudly in one ear than it is in the other. Although this difference is very slight, the brain can identify it. If a sound is to the right, the right ear will receive it a fraction of a second before the left, and it will also receive it a little more loudly.

Many animals are able to improve on this ability by moving their ear pinnas in all directions, making fine adjustments as far as direction is concerned. Those animals that, like man, cannot move their ears often move their heads in a wide arc. Birds do this, and the owl is even able to turn its head 180 degrees each way, thus completing a circle.

Aside from the fact that they have no external ears, fish cannot easily judge sound position because this judgment is more difficult in water, due to the greater speed of sound in water and the consequent lengthening of its wavelength. Judging direction by the difference in the loudness of a sound between the two ears is most efficient with high frequencies, especially when the wavelength of the sound is shorter than the width of the head—that is, when the obstruction of the head to the sound reaching the second ear is at its greatest.

Nature has provided many birds with protection against hawks and other predators. These hunted birds fade their warning calls from loud to soft or run them down the scale. Some birds alternate notes, loud and soft, so they cannot be pinpointed precisely by predators. But they do not do this when calling to their own species.

The human ear can withstand a range of volume or loudness from a point where a sound can only just be heard when all else is silent (called the *threshold* of hearing) to an amplification of ten thousand million times that level. But such an intense volume can produce permanent damage to the delicate ear nerves if it is prolonged. We are not sure just how much volume other animals can stand. Although this must vary greatly, some animals on which observations have been made do not seem to be bothered by a degree of loudness that is uncomfortable for us. For instance, some of the underwater calls of seals can be very uncomfortable for skin divers, but they do not bother other seals.

The large ear pinnas possessed by nocturnal animals has already been mentioned, but it must be emphasized too that large ears alone do not guarantee the appreciation of a wide range of sounds or volume. They only serve to pick up and direct the greatest possible amount of sound into the middle and inner ear chambers from all directions. Such animals must also have a very great area of the brain devoted to interpreting what they hear. This applies especially to animals that use echolocation or sonar, in which sounds are transmitted by the animal and the reflected echoes are interpreted so accurately that even in complete darkness the smallest creatures can be detected and captured.

It is easy to understand that the hearing range of any animal must be most sensitive to the sounds made by its own species. Man and most birds have rather limited ranges of sensitivity—especially man—when compared with bats and some whales. Some insects can hear frequencies as high as 100,000 vibrations (cycles) a second (shown as cps.), and it has been claimed that the limit is probably more like 300,000 cps. for some insects, but that has yet to be proved.

Dolphins appear to hear sounds between 20 and 179,000 cps., and they can also transmit sounds within that range. Not all measurements on dolphins agree, however. Tests carried out in the United States have shown definite hearing up to 153,000 cps. and certainly very positive hearing at 120,000 cps. These same animals seemed to be frightened away by sounds below 400 cps., and this may be the kind of sounds made by some of their enemies. It has been found too that the larger whales are very sensitive to asdic (*A*nti-*S*ubmarine *D*etection *I*nvestigation *C*ommittee, or ship sonar) frequencies, which range between 20 and 40,000 cps.

The dog and the rat can hear frequencies up to 40,000 cps. (some experts say 70,000 cps.). One of the large green grasshoppers [108] can hear up to 100,000 cps., the crocodiles from 50 to 4,000 cps., and some sharks can only manage to respond to vibrations from 100 to 1,500 cps. This is a very poor range when compared to other animals; even man can reach 10,000 cps.,—and sometimes higher than 20,000. Although man has a much lower sensitivity than many animals, he has all that he needs, and above all he has a level of *understanding* or *comprehension* of what might happen in a given set of circumstances that few other animals possess.

2. The Ways Animals Make Their Sounds

There are few living creatures that do not produce sound in one way or another—especially those that live on land. All sounds made by a certain species have meaning for other members of that species.

What Is Sound? Anything that vibrates and anything that rubs against something else creates sound. Even the gentle passage of air through a rock hole, water over stones, or stones rolling on stones create sound by friction. It seems natural that while much of the animal world has learned to produce sound vibrations with some kind of air passage. many other creatures use rough friction, like rolling stones. This means that many kinds of sound receivers had to have evolved at the same time.

Sound moves away from its source through a medium—air, water, or solid material—by vibrating the particles of the medium without moving any part of the medium itself. In water, for instance, the particles of water are moved back and forth, but no current is created. When a particle moves away from the sound source, this is called *compression,* and when it returns to its previous position to complete a vibration, this is called *decompression.* If we illustrate these concepts on a graph, a wave pattern is drawn showing peaks and troughs. The peaks represent the compressions and the troughs represent the decompressions. The distance from one peak to the next is the *wavelength,* and the number of peaks (vibrations) that pass a given point in a second is known as the *frequency,* represented as cycles per second (cps.).

These concepts are important to remember as we discuss investigations into sounds used by animals for communication. The height of the peaks on our graph represents the loudness of, or the amount of energy in, a sound. This loudness is also called *intensity* or *amplitude.* A factor that depends on both frequency and intensity is called *quality* or

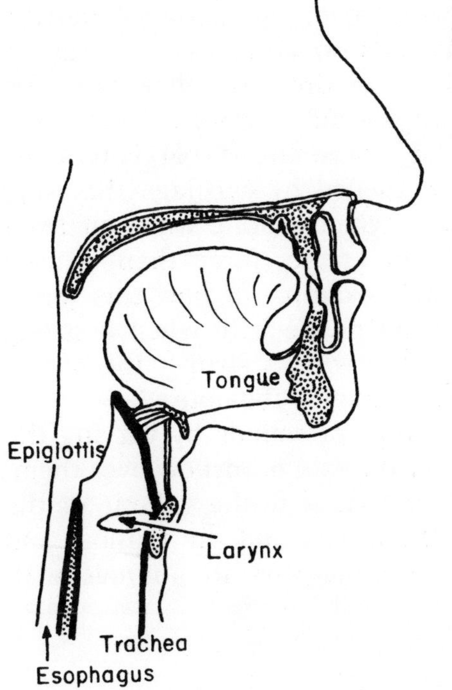

Figure 3. The human *larynx* is in the windpipe (*trachea*) from the throat to the lungs. It is the same in all land mammals.

purity. But we are more concerned with wavelength and frequency at this point.

Sound travels at about 750 mph. in air at sea level, but it travels much faster in denser media like water and ground. It is also louder in denser media too.

Sound-Producing Instruments. Vibration waves in the atmosphere or in any other medium can be produced by voice, drumming, friction, or striking part of the body, such as a hand or hoof, against an object or the ground. All these ways may carry some significance. Most land vertebrates have voices, which are made possible in most animals by the use of a *larynx.* To make their sounds, birds use a *syrinx* and fish use the gas bladder, pharyngeal bones in the throat, spines, fins, or jaws, while insects vibrate abdominal plates or rub together special limb formations to produce *stridulation.*

The great difference between mammalian voices and the much higher pitch of bird calls suggests that the larynx and syrinx are different in structure, and they are. The larynx is a chamber at the upper end of the trachea (windpipe), just beyond the point where the throat enters the trachea. The larynx is strengthened by cartilage that supports a number of *vocal cords,* which vibrate when air from the lungs passes over them in greater force than is used for normal breathing. A simple example of how this works is seen in the vibrating of a telephone wire when a strong wind is blowing. The tighter the wire is strung, the higher the note; the slacker the wire, the lower the note.

Larynxes produce a wide range of sounds by varying the tension of the vocal cords and the rate of airflow over them. The size of the chamber is important to the volume of the sound created. The size of the larynx and the number and length of the vocal cords naturally vary in animals with

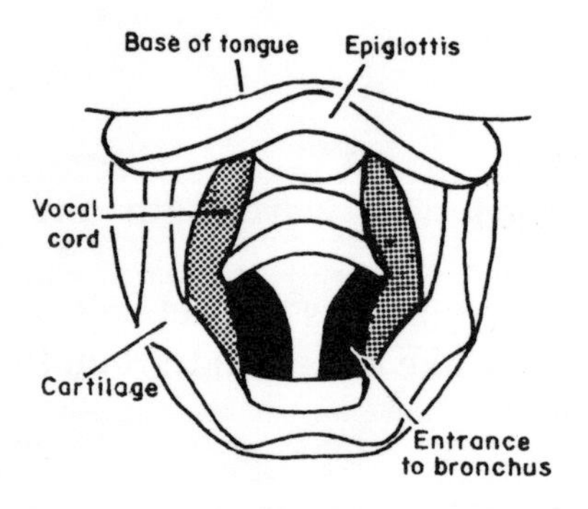

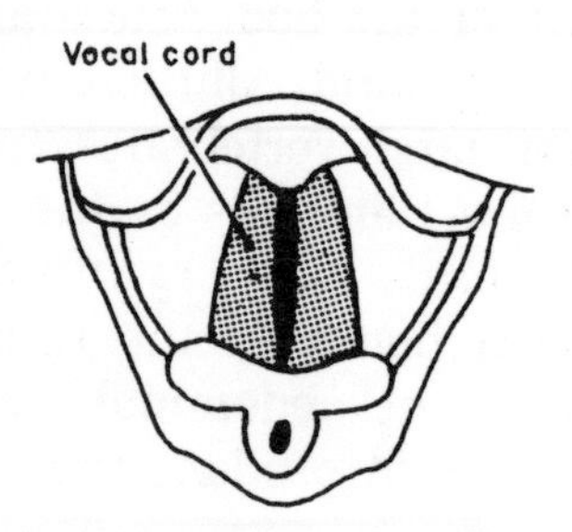

Figure 4. The vocal cords in the larynx operate by changes of tension and of the space between them. In the upper figure they are wide open, as seen when looking down from the throat toward the lungs, and in the lower figure they have been closed by muscular action.

the range of sound frequencies a species uses. The throat formation and the position of the root of the tongue also differ. This is important to voice quality in man and some higher mammals, for the movement of the tongue is part of sound formation. Perhaps most important is the changing of tension in the vocal cords by several small muscles. As the cords are drawn tighter, the note (frequency) rises.

Examination of all animal larynxes shows that they are simpler in animals that have the shortest range of sounds and more complex in animals that can make a wide range of sounds. But the complexity of the larynx is not always so important as throat formation and the position of the tongue, which make it impossible for great apes to talk like men. The apes can make sounds, but they cannot form these sounds into words. This will be discussed further in Chapter 5.

There are modifications for amplifying sound in the larynxes of some mammals. For instance, the hammerhead bat [55] has an immense chamber to amplify its signals. The sea elephant [79] uses its trunklike nose as a resonator to amplify its voice, and some other mammals, birds, and many amphibians have throat pouches for the same purpose.

3. Understanding Messages Through Vision

If we wave a hand to a friend across a field or park, he can pick up a great deal of information from this movement, depending on how vigorously we wave, how high we hold our hand, how we bend our arm, and for how long we wave —even on how we stand while waving. Our friend will read a definite message in our gesture. It is the same way among animals. For some animals, movement and gestures are the only kind of language used to a great extent, so they need keen vision.

Visible signals may have been among the first forms of communication developed in the animal world, but we cannot be entirely sure of this even though the earliest types of fish show signs of having had well-developed eyes. Certainly light began to affect living cells many millions of years before sound waves did, for we find the beginnings of eyes in some of the most primitive organisms. There is even an eyespot in a single-celled creature, the *euglena,* but it is so simple that it provides nothing more than the ability to tell whether it is light or dark.

Much more complex organs of vision became necessary before actual patterns and shapes could be recognized. But the fact that vision was one of the first senses to develop does suggest that communication through signs, color, movement, or some other visible means must also have developed before sound or odor.

Combining Senses. It was inevitable that once several senses had evolved nature found it necessary to coordinate these —even when one was used more than the others—and eventually animals became able to use one sense to check what they experienced with another—a kind of strengthening of impressions. We can use ourselves to illustrate this idea.

If we hear a voice coming from a group of people we do not know, we may try to guess who is making the sound. But we will not be sure unless we see a mouth move at the same time as we hear the sounds. Seeing the mouth move will identify the speaker, and we will get both the message and its origin. In other words, we coordinate vision with hearing to gain a complete understanding of the situation. Our vision checks or confirms our hearing.

There are advantages and disadvantages in using vision alone to receive communication signals. The main advantage in the wild is that an animal does not betray its position to its enemies as it would with sound. Another advantage is the ability to recognize signals that rely on color and to judge distance more accurately than with any other sense except for echolocation.

The disadvantages of using vision alone to receive communication are the restricted number and speed of signals possible, the difficulty of communicating at night without special light organs (such as a number of fish and insects have), and the limited range of the signals if there are trees or other growth in the area. Powerful sound signals, on the other hand, will carry over great distance.

Perhaps the only animals that can use visible signals over relatively great distances are sea birds. They have no trees, hills, or other obstructions to their vision, and their vision is many times more sensitive than our own or that of any other animal.

Most visible signals must depend on the standing attitude of an animal—the position and movement of its head, ears, and tail—and on sudden movement or even the absence of movement when the animal is tense. Not only are these signals understood between members of the same species, but animals that are hunted also learn to some extent to interpret the moods and the meanings of the movements of predators that are likely to hunt them. For instance, when a lioness separates from a pride and begins to move in a certain manner, wildebeests, zebras, and a number of other grazing animals become alert, sensing that the lioness is about to hunt.

Among the primates at least—and perhaps among other animals in which we do not recognize it so easily—facial expression is a signal. We see it to some extent in dogs, but it is among the apes and man that facial expression is in its most highly developed form. Apes and humans clearly show anger, desire, friendliness, fear, curiosity, hostility, affection, and many other feelings. All are significant messages, part of a silent language for those who need to be informed.

Color Change. When someone is angry, his face flushes. When he is frightened, his face becomes pale. This facial change is the result of increased or decreased blood to the face, and it is due to emotion. Many animals change their color as a result of emotion, as a result of hormones in the bloodstream, and when trying to frighten away an enemy. In their way these signals are also a form of language.

The change of color due to hormones is usually seen during courtship and preparation for breeding, although it is often used for challenging rivals and enemies. It is seen in a number of species of all groups of animals from fish to mammals. In the case of courtship it is a message from a male to a female or vice versa that says: "It's spring; follow me." These animals cannot say anything like this with words, so they say it in the only way the species can—by color change.

4. *Using Odors as Messages*

Odor consists of particles of a substance carried to a nose or other organ of smell by currents of air or water. If the particles are sufficiently strong or in sufficient numbers to irritate the sensitive membrane of the nose, a message will be sent to the brain, where the substance will be recognized. Some animals' organs of smell are so sensitive that identification of the species and of individuals—even of the sex of an individual—by odor is used to some extent in all groups of animals. It is also used for signaling, for warning, and for attraction.

All odors are produced by glands, and some odors give a special character to the body's waste products (excretions). Odors used to attract the opposite sex are called *pheromones,* and their chemistry is exceedingly complex because they must carry the signatures of the species, the individual animal, and sometimes the small group or family to which the individual belongs. We do this with our eyes. We may look at a man and see that he is an Indian, that he has certain kinds of features, and that those features are those of a family we know, and to which he obviously belongs.

Although odors are the most persistent form of signal, most of them still have a very limited life. When carried by the wind they can travel considerable distances in a short time. One great disadvantage of these chemical signals is that they cannot be changed or canceled in the way that visual or sound signals can. Once they are launched, nothing further can be done about it. Like rockets they cannot be brought back.

Odor Glands. Special odors can be secreted by almost any part of the body—the face (see Figure 9), the chest, the rump, the arms, the legs, the back, and so on. Ants, for instance, have several different kinds of odor glands in different parts of their bodies, and each odor is used for a different purpose. Some animals secrete special odors with their urine, using it like any glandular secretion to mark their territory, to be recognized, to create a trail, or to mark a mate (see Chapter 6).

Even man has his individual glandular odors. Most of them are in the skin, but we have lost the ability to distinguish them one from another. In fact, our culture has come to consider odors objectionable, and we are encouraged to use all means possible to remove them or cover them up. But a dog can still separate us easily by our odors.

Although odor glands may differ in location, form, size, and the strength of their secretion, they are all simple groups of special secreting cells producing a constant odor-carrying fluid that never varies. Nothing is ever without its exception, however, and there appear to be amphibians that give off odors without any special glands to manufacture them. In these, just the odor of the skin is responsible.

Some odor-producing glands act by manufacturing a substance that can give off its odor only when attacked by bacteria, while others come into full function only when reached by hormones carried by the bloodstream from other internal glands. These are the emotional glands, and they can be gradual in their action (as in the breeding season) or they can be virtually instantaneous (as in anger or fright). These may hardly seem to be signals, but they are. Within a close family group they can be very definite signals, because survival may depend on anger or alarm being transmitted from one member of the group to the rest, so members will combine to fight or run to safety. The use of the signal in this way is quite clearly seen in ants.

Odor-Receiving Organs. The use of odors for communication or tracking demands very sensitive noses or other organs of smell capable of picking them up. Some of the odor signals used for marking out territory limits, for branding mates, and for laying trails hold their strength for fairly long periods, but there are short-lived odors used as warning signals within a close group and as attack signals. All of these signals give the greatest amount of information to those animals with the most sensitive noses. This can be most important when picking up a fast-fading signal that marks the path to a food store discovered by another member of the family group.

Odors carried by a breeze or a water current will have their range increased, but only in the direction of the movement. Without a breeze or current they will spread in all directions, but will lose their strength in a shorter distance. It is obvious that no matter how sensitive a nose may be, there will be times when an odor will be carried away from it if the odor is only released into the air. This suggests that laying a trail for another family member may be a better signal, but this is not always possible.

Releasing an odor into the air is probably the most effective method of communication for insects that do not fly, because of the limited distance they can see. These insects are so close to the ground that their horizons may be only a foot or so away from them. Odors will not only travel farther than these animals can see, but in most cases farther than they can hear too. These insects have thus developed extremely sensitive odor receptors that can identify just a few particles in a million particles of air. The same can be said for many marine animals.

The sense of smell is experienced in a variety of ways: by a nose in vertebrates, sometimes by surface cells on the body as well as a nose in fish, and by special sensory cells, or *sensillas,* in insects. Sensillas are situated on the insects' antennae, and they are discussed at length in Chapter 13.

A nose, whether it has the customary pair of nostrils of most vertebrates or the single one of the lamprey, always works on the same principle. A chamber is lined with specially sensitive cells, situated so air or water passes over them and allows them to absorb the odor particles it carries. In air these cells pick up more particles with each sniff or breath, so the odor builds up in the receptor cells and becomes stronger. Each cell is connected by a nerve fiber with the brain's smell (olfactory) center, which is more developed in animals that use smell as the dominant sense than in animals that rely on vision or hearing.

There are many odors in nature, and they all mingle. In order to recognize an individual signal odor, all the other odors must be filtered out or masked in the brain, unless the signal wanted is so strong that it overwhelms all the others. This masking out of unwanted odors is just as important as the need to mask out unwanted sounds when using the sense of hearing, but just how it is done is not yet certain.

PART II

COMMUNICATION IN MAMMALS

5. *How Primates Communicate*

Because the highest apes seem so much like man, their behavior has been carefully observed for a long time. Some people have tried to teach them human skills, sometimes with fair success. But although apes have complex languages of their own, they have been unable to copy more than a word or two of human language and then only with sounds that are somewhat similar, never the same. One chimpanzee was raised in a family that intended to teach it to speak if possible. The chimpanzee was able to learn only four words in six years, and these words were not very clear. Efforts to teach apes to speak must always fail, not because an ape is incapable of learning but because the root of its tongue is in a different position from that of man's, and an ape cannot use its tongue the same way that man can.

Chimpanzees. At least one chimpanzee has been taught to converse with its trainer in a simple way by using special plastic symbols and letters. It can identify objects and their colors and give instructions for whatever it wishes its trainer to do with them. At the end of two years' training this animal could use 120 words. Another chimpanzee trained in American sign language learned 130 words in signs in four years. This is comparable with the ability of some human children, so chimpanzees are obviously capable of subtle language expression, even though their tongues and throat structure are unsuitable for making words.

Although neither of these chimpanzees could say a word, they could express their ideas in other ways. This is of course what they would do in their natural environment. Chimpanzees have a wide range of calls, posture signals, and facial expressions with which they communicate, just as we use nods, hand gestures, and eyebrow movements to add to our word meanings. These movements were used by man long before he used words, and he has never given up using them in spite of his use of word language. When an ape makes a gesture, however, it may not mean the same thing as when a human being makes the same gesture.

A chimpanzee's facial expressions cover a wide range of feelings, such as relaxation, alertness, aggressiveness, threat, attack, fear, threat with fear, hunger, pain, play, and so on. Most of these feelings also produce voice sounds—some of them very powerful; yet there are times when, like man, the chimpanzee prefers to be silent. If he thinks a fight is coming and he doesn't feel like fighting, he is quite likely to sneak off into the jungle without a sound.

These animals express contentment with low grunts when they are eating, and similar grunts are heard when they play. A low-pitched "hoo" will be used to greet another chimpanzee, and occasionally friends will fling their arms around each other when they meet. Sometimes they will just touch hands or grin, or they might give several panting grunts. So much depends on the circumstances and how the two animals feel about each other.

Greetings between two large groups can often build up to loud, excited calls and screams. At the approach of a hunting animal like a leopard or when they are really angry at the approach of another chimpanzee, their calls can be loud, defiant, repeated, and a bit hair-raising. Any juveniles in the area will be giving their own short, sharp screams of fear. When a male leader is about to cross a ridge or a hill, he gives a series of hoots followed by three or four roars to let any groups on the other side know he's coming with his family. This makes certain that their sudden appearance doesn't start a fight because they have surprised the others.

Chimpanzees can be very noisy during their dance sessions. They thump, clap, stamp, scream, tear up vegetation, and even beat on hollow logs or tree trunks. They sway and dance, and the juveniles join in as though they were enjoying a human "rock session." The animals may keep it up for half an hour, their laughter making a kind of panting sound.

Chimpanzees have air sacs in their throats for amplifying sound, and when a chimp grunts, it is thought to be due to the release of air from the animal's air sac. Such air sacs are found in a few species of all other orders. The largest sacs of all primates are found in orangutans. Macaques, baboons, and a number of other monkeys have them, but lemurs do not. None of the sacs are as large as the orangutan's, although the gibbons seem to make tremendous use of theirs.

Quite often chimpanzees will express themselves only with their faces or by contact. The greeting of a juvenile can be a forward protrusion of the lips, while that of an adult might just be a broad grin. Chimps will crouch, grin, smack their lips, chatter their teeth, and pout for various reasons, and when frightened they will draw back their lips and show their teeth with an open mouth. Showing the teeth both with and without sound is common to all apes, and there are many ways of doing this—all with different meanings.

If one chimp approaches another one rather nervously, it will make its mouth into a small opening and raise its eyebrows. Then it will touch the other's hand in greeting. If an ape is frustrated, it will open its mouth and project its lips with great force; but real anger brings the hair on its head upright and bristling.

Contact signals are also very prominent among chimpanzees. If they are nervous, they will reach out and touch each other for reassurance. The first one reaches out palm up, and the other accepts the palm with its palm down. A dominant male will touch the back of another male to give it confidence. A mother will touch her young when she is about to move away, or she will tap a tree trunk when she wants junior to come down to the ground.

A chimpanzee will pat a branch as an invitation to sit

alongside him. If he wants to sample another's food, he will hold out his hand palm up. All these silent signals are as useful and expressive as speech, and they serve the same purpose at times when (unless the group is large and powerful) noise will attract the unwanted attentions of the big hunting cats.

Gorillas. These largest of primates may seem to have a smaller vocabulary of expressions and sounds than chimpanzees, but much of their vocabulary can be recognized for what it means, even though it is different from that of the chimpanzees. Gorillas seem to use gestures as much as sounds. About twenty-two communication sounds have been identified in mountain gorillas.

Besides voice sounds the most mature males indulge in chest-beating not to express anger, as is so often assumed, but as an alarm signal or as a release of tension. Ground-thumping is another kind of tension release; another alarm signal is a hooting kind of bark, which probably conveys a different kind of message from chest-beating because it can also signal curiosity.

The screams and roars of an angry gorilla are probably terrifying to all other animals in the area, but the same gorilla can sound quite gentle when with three croaking grunts he tells his group that he has found food or when he wants to express contentment. What sounds only like grunting to us may have slightly different tones for different meanings, because grunting is used for several purposes, and we cannot easily hear the differences that make this possible. At times gorillas will grunt continuously, evidently to keep the various members of a group in touch with each other.

A gorilla will demonstrate friendliness silently by folding its arms across its chest, similar to one man bowing to another. A sign that a group is very contented is a kind of belching. When one animal starts belching, the whole group will join in. This sound is also made by orangutans by inflating the air sacs under their chins, but when the orangutan does this, he means something quite different. He means that he is prepared to fight any intruder that may wander into his territory.

When resting or feeding contentedly gorillas are likely to be heard quietly purring, grunting, humming, or even making what we would probably think of as grumbling sounds. Most of their sounds are low-pitched anyway, between 100 and 2,000 cps. But a kind of sharp, harsh grunt by a dominant male will snap his whole troop to attention. A female will discipline her young in the same way. A couple of short, sharp grunts by a group leader will mean: "Look out; danger." Females and young will then move in closer to him.

Orangutans. Although orangutans belch when they are ready to fight, these animals are ordinarily very peaceful. They are also very solitary. Since they do not live in groups, they have very little need to communicate, so they have never learned to make many sounds. They are perhaps the quietest of all apes. For the same reason, their faces change expression very little. Orangutans may not even have all the feelings and emotions that chimpanzees and gorillas have, although they will of course experience the simple ones of fear, hunger, and defiance of enemies.

The Rhesus and Similar Monkeys. The rhesus monkey has been used extensively for laboratory experiments, and it has been found that this animal also has quite an extensive vocabulary, including screams, screeches, roars, squeaks, shrill barks, and pants. These sounds have been analyzed, and they show that a confident monkey threatening another will roar loudly, but if he's not sure of himself and needs moral support, he will break this sound up into short notes evoked in a series. If he has practically no confidence at all, he will give just a single bark, and if he's losing a fight, he will scream. If the fight is lost and he's had the wind knocked out of him, he will give a series of high-pitched squeaks. This is like a human being saying, "Okay! Okay! I give in."

If a rhesus male has the courage to challenge a member

of the group that ranks higher than he does, he will do this by screeching, the pitch rising and falling—a sound that he also uses when he is excited. When another animal threatens him, he screeches this way, but he breaks it up into short screeches called a "geekering" screech. As among humans a female will often try to pacify the quarrelers and reduce their aggressiveness by showing some affection or by trying to distract their attention.

While a shrill bark is an alarm call at any time, the rhesus seems to have six different sounds for when it senses the presence of a dangerous hunting animal; it has different calls for different kinds of predators. It even has a different call for a perched eagle than for a flying eagle. Each call produces a different response or action in those that hear it.

Grooming is a constant monkey pastime during rest periods, and even this has its language. When one monkey goes toward another to groom its fur, it smacks its lips rapidly and sometimes sticks out its tongue between each smack. The other responds to this by relaxing. A female will go toward a nervous baby with the same lip-smacking to make it feel comfortable, and it seems that this gesture and sound are very important messages.

Lip-smacking is also part of baboon language, and it means "I want to be friends" or "I want to join your group." But baboon language seems to be more precise than the language of rhesus monkeys; certainly the gestures seem more definite. Then too the calls of baboons are either loud or very soft; baboons seem to have no ability to graduate them.

To make sure that a tribe is well spread out and that an area is not overpopulated, careful spacing of the monkey groups is necessary. To ensure this, the black-and-white colobus monkey [32] gives off a roar that can be heard for just about a mile. Howler monkeys [7] do the same thing. When other groups hear this roar, they stay just out of reach.

A monkey that has a large number of visible, sound, and odor signals is the titi.[24] The visible signals are fairly simple, but the sound signals are not. And the titi uses its nose much more than most of the monkeys found outside the American continent because adult titi males have an odor gland in the center of the chest and they mark the edge of their territory by rubbing this odor on branches.

Because these monkeys have tails, they are able to make signals that tailless apes are unable to make. Titis can express their feelings with their tails by intertwining them with each other for close contact, rather like a boy and a girl putting their arms around each other. Titis make many sounds too: whistles, chirps, grunts, moans, gnashing, and pumping noises, probably all of them unfriendly, because they go with hair standing up, tail lashing, and arching the back; and when predators are around, sounds are combined with swaying steadily from side to side.

We tend to raise our voices in an argument or when angry and lower them when we feel relaxed. Monkeys do exactly the same thing. A particular sound made very loud will mean something quite different from the same sound made softly. A call can be used to express fright or, with a slightly different tone, loneliness if the animal is lost. A lost monkey may be both lonely and frightened, and the tone of its call

Figure 5. Grooming is a constant occupation of monkeys, and it has its special language.

tells us if it is more lonely than frightened or more frightened than lonely. It may use yet another tone if it is raining and feels cold and uncomfortable. The purpose of such calls is always to tell the story to other members of the group or to attract them for company; otherwise the calls would have little purpose.

When titi monkeys have finished a family quarrel, some continue to moan and grumble for a while, just as humans do when one member of the group must have the last word to help him feel that he is the winner or to show that he is not quite ready to forgive.

Activity at night makes it impossible to use many visible signals, and most night contact must be made by sound signals over distances that are not very great, so most of the sounds of the night monkey [15] are quiet. This monkey has nine or ten distinctly different notes and calls; most other primates have at least this many, with various changes of tone to give different meanings.

Lemurs. The lowest members of the primate order are the lemurs. Their faces show little expression, and they are not as intelligent as monkeys. Their vision is not quite as sharp either, although it is enough for their needs, and their voice sounds are less developed. The ring-tailed lemur [60] seems to have just a few main sounds around which it builds its whole range. Two of them are sudden: a shriek for alarm or danger and a howl given by one of a troop both at dusk and around midday before the troop takes its siesta. The howl may be to bring the troop close together.

Another kind of howl is used to tell other troops where they are; this may also be used by one lemur to tell his family where he is. Before a troop moves to new ground, it sounds a series of clicks, as though calling everyone to attention before marching. This series of clicks is often greeted with moans and wails, like the objection we would get from young children in the same circumstances. In the breeding season males bark quite a bit, and the rest of the troop answers with catlike meows.

The ring-tailed lemur has musk glands on the inside of each of its arms. When it is going to fight, it drags its long, bushy tail over these glands to give the tail an odor; then it waves its tail above its back to create a breeze with a stink or to launch this smell on any breeze there may be. When lemurs are on the ground, they carry their tails erect. It almost looks as though they were trying to outsmell each other.

R. G. Busnel, a scientist, has made many observations on primates. He has decided that the range of sounds used by the animals in this order is from 269 to 12,150 cps., but it is possible that some go lower than 269 cps., and probably very few ever go as high as 12,000 cps.

6. How Other Mammals Communicate

Very few mammals other than apes and humans have a wide range of facial expressions. Other mammals can not move their face muscles to the same extent, but they have other visible signals besides the sounds they make. These lower orders also use odor to a much greater extent. Just a few ways in which visible signals are made to other members of their species include arching the back, lowering the head, pawing the ground, moving the tail, and contracting the body size.

The tail is an important organ of communication in some species, but the number of signals that can be made with it is not very great because there are only certain things that can be done with a tail. The number of sounds that can be made by lower animals is limited too, but messages that combine sounds with visible signals provide a much larger range of communication. However these sounds and signals are used, they are always adequate for the species concerned.

When sounds are used by mammals for signaling, they can probably be grouped as messages for:

1. *Survival.* These would have meanings of warning, alarm, threat, aggression, and rivalry.
2. *Breeding.* Finding a mate, actual mating, communication over a distance, and close communication.
3. *Rearing.* Warning, affection, contact and reassurance, and calling young to the parents or the group.
4. *Social life.* Alarm, aggression, pain, fear, excitement, comfort, close contact, and the feelings of the entire group.

Canines. Because we live so much with dogs and understand them and because they will obey us better than any other animal, many people think that dogs are more intelligent than other animals. But this is not always true. Because dogs are friendly animals that enjoy company, they have accepted man's ways and adapted to them. As much as we

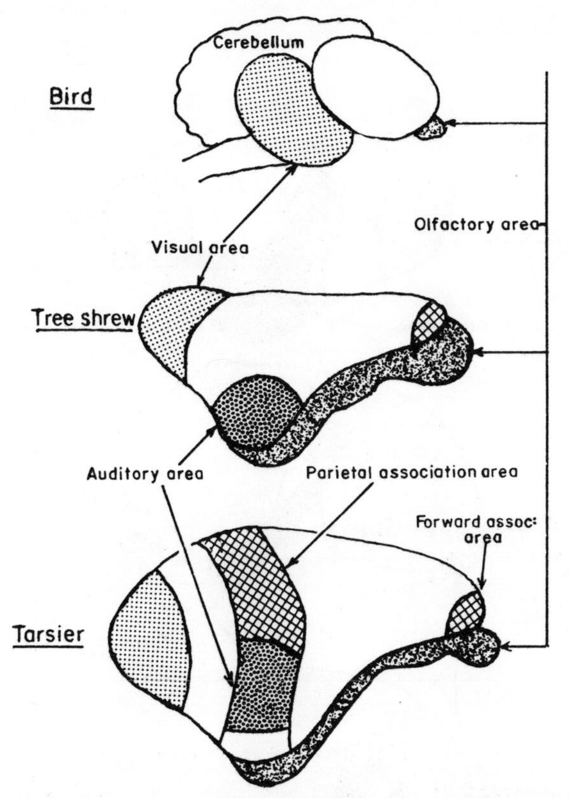

Figure 6. The visual center in a bird's brain is almost as large as its other sense areas combined, but a bird has a very small center for smell. By contrast an insectivore (tree shrew) and a nocturnal primate (tarsier) has greatly developed smell areas and relatively small visual centers. Forward and parietal association areas coordinate movement and experience with the senses.

understand dogs, it is doubtful if many people have even noticed the subtle differences in their behavior in different situations. For instance, dogs have a different bark when challenging humans than when challenging other dogs, and their body signals have slight differences that few people notice.

Dogs seem to have about six kinds of voice signals, which include greeting members of their family, playing among themselves, chasing wild animals, and teasing other ani-

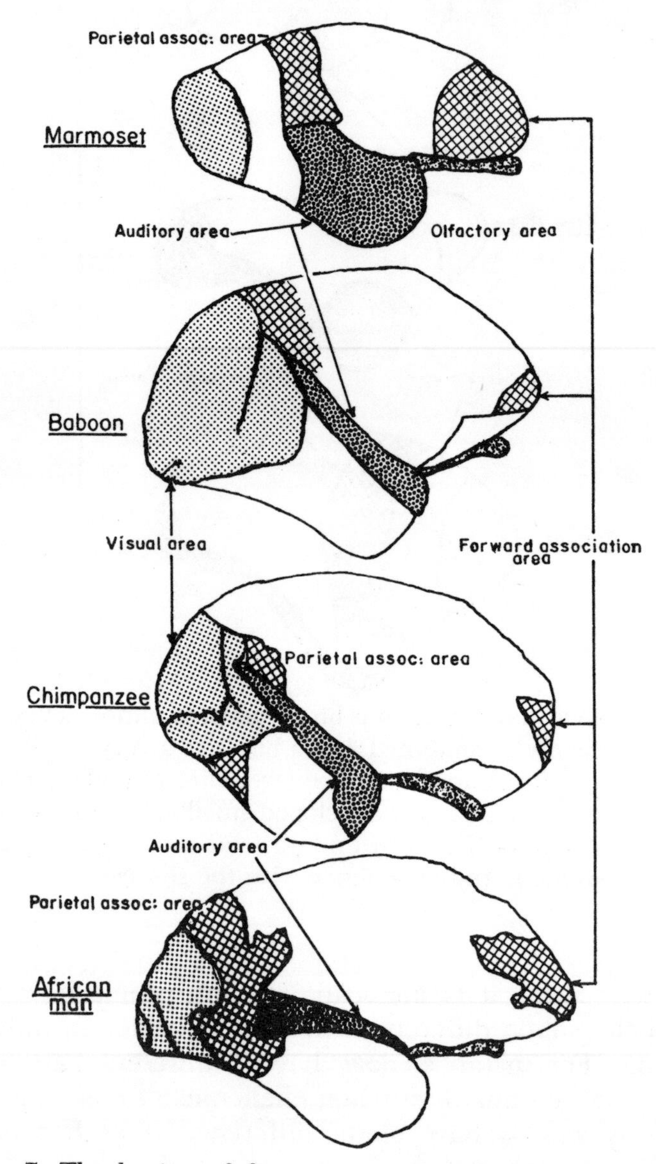

Figure 7. The brains of four non-nocturnal primates including man. The brain enlarges greatly as we go up the evolutionary scale. The centers for smell get smaller, and the visual centers and association areas increase in importance. *The brains are not drawn to scale.*

mals. Not all teasing is noisy; females do not always make sounds when they playfully nip males, which will accept this more patiently than they would if other males attempted to nip them. In the six different voice signals, there are many slight tonal differences that we do not usually hear, but they mean a great deal to other dogs. The differences in tone are almost like conversation.

Although domestic dogs are more or less solitary animals because they belong to human families, they really are very social and would like to have their own family group too. But this is usually possible only on a farm or estate where a number of dogs may be needed. The need to be social applies to all dog types. The fox likes to be one of a family group as well, although not of a human family. It will cry if it is lonely, and the more lonely it feels, the more it will show it by letting its tail droop lower and lower. The fox will yap with pleasure like a dog when it is playful. It has many other sound signals, and although these sound like those of a dog, they do not always mean the same things.

Wolves behave in much the same ways as dogs in some things, but they are able to live more in groups than domestic dogs. The coyote will sometimes sit alone and howl, but like other species of wolves it also enjoys communal howling. For coyotes, howling may be like our singing in church. Some canine species have special sounds for their breeding season, and this applies to both males and females. Other kinds of mammals also have special sounds for the breeding season—especially the hoofed animals, including the deer.

The hyena's laughter has nothing whatever to do with fun or humor; it is not laughter at all. It is an eating signal, perhaps calling the pack to a food-find or to a prey that has been killed. If hyenas are disturbed by the approach of a predator such as a lion, they give an alarm signal of very sharp, deep grunts to warn other nearby hyenas. Even within a group of animals as closely related as all the dog types, their individual signals may not be the same for any particular situation.

One of a dog's most important methods of signaling is with its tail. A tail well up shows confidence or maybe readiness to fight; a tail well down shows the opposite—fear. A dog also shows fear by laying its ears back. Wagging the tail sideways shows relaxation, pleasure, or anticipation; and the speed of wagging tells how strongly the dog feels that emotion. But wagging a tail up and down is said to be a sign that the dog knows it has done something wrong and may be afraid of being punished. It may be trying to say, "Let's make up." When a cat swishes its tail from side to side, it is for the very opposite reason that a dog does it—often because it is angry.

When a dog shows the whites of its eyes and keeps its tail motionless, this can be a sign that it is going to attack. Drawing back the lips is also a sign of possible attack; it is certainly a threat. In almost all mammals a steady stare is considered to be a threat, and when a dog is doing this it can be goaded into attack by staring steadily back at it. Moving the head away usually breaks the threat.

The variety and meanings of tail signals and stiffness or looseness of the legs and body are quite numerous. Complete rigidness of the body with bristling hair would never escape our notice, and we would never tease such a dog or even approach it. But it would be just as important for us to recognize slightly less rigidness even though its meaning might be different—especially if the lips are drawn back or the ears are not up.

When a dog attacks another animal, it makes a good deal of noise. This is not just to frighten the other animal. The ancestors of dogs lived and hunted in packs, and even now some of the noise a dog makes when fighting is to let other members of a group know that he is protecting them. Some of the noise is possibly just for the benefit of any other dogs just looking on, if the fighter likes to "show off," but much depends on just how bad-tempered the dog may be.

We must not forget how odor is used by dogs and many other mammals—odors that carry the brand of the species and also a personal signature. When any canine—whether it is a dog, a wolf, or a fox—raises its leg against a tree, it is saying with its personal odor, "This is my territory; keep out." When another dog follows and does the same thing, he may be saying, "No, it's not; it's mine." In the wild state canines mark out the boundaries of their entire hunting ground with the odor of their urine, and this is usually respected by all other groups and individuals of the same species.

Perhaps most information is given by the urine signals of female canines. Their urine odor varies year round according to their readiness to mate. R. H. Smythe, a British veterinarian, has claimed that there are fifty-two different variations in the odor of the female urine in a year. A dog can get information from each variation to the extent that he can calculate the exact day when a female will be ready for mating, and dogs from miles around will congregate in her home area and wait for her on that day.

Small glands around the anus of every canine identify the sex of the animal in its excretion. In this way any canine can tell which other animal has passed by a given spot and probably when, because of the strength (age) of the odor. Thus, all canines gather an astonishing amount of information through their noses.

Marking the boundaries of territory is also practiced by badgers, martens, mongooses, and agoutis. Like dogs, brown bears use their urine to mark their boundary, and so do bison, house mice, brown rats, and many other animals. European rabbits, cavies, agoutis, and porcupines also mark their mates with their urine, while hamsters and foxes will encircle a female's territory with urine, as well as their own.

Some Ungulates (hoofed mammals). Horses' ears are probably as sensitive as dogs', but they use sound much less. In the wild they use their ears to get information from the sounds of hooves and from the sounds made by other animals. It is the same with wild cattle such as bison, in which communication within a group is by body position, stamping, body-stiffening, head-swinging, snorting, and bellowing. The wildebeest (gnu) snorts too and holds its head high to signal a warning when a hunting animal such as a lion approaches. A number of deer signal this kind of warning with their tails.

Figure 8. Ungulates have very flexible and well-controlled ear pinnas, which can be turned in many directions to pick up sound and identify its location. In many animals pinnas are also used as signaling devices.

Perhaps in all the ungulates the simplest and yet most definite of their signals are those between females and their young; these signals are often by touch alone. A mare, for instance, may convey the same message to her foal by rubbing her head along the foal's neck and blowing a little air through her nose as a human mother would do to her small child by placing her hand behind its head and saying a few quiet, kind words.

Elephants use many sound signals—trumpeting, snorting, growling, squealing, and roaring—all for different reasons or situations. In this way they express greeting, intention to attack, warning to other elephants, distress, and call their young. In fact, the elephant's range of sound messages is very extensive for an ungulate.

The elephant's sense of smell is quite phenomenal, as one would expect for an animal with such an extended nose, so it is not surprising that it has glands on the face that can be used for marking territory. These glands appear to be most active in the breeding season, however, so they may be used most for branding mates and perhaps they tell other elephants only that a female belongs to a certain male. The female and her mate will also use touch and visible signals in their rather gentle courtship.

Facial glands are prominent on antelopes (see Figure 9), and these too are used both for marking territory and for branding mates. Antelopes mark territory on twigs, branches, or leaves on a level with their eyes. In this way the odor can not be missed by other members of their species. Many ungulates also have glands close to their hooves, and these glands leave a trail for others to follow. This keeps a group together when visibility is poor. These particular odors do not seem to be detected by predators which is very surprising. It suggests that there may be a very limited range of smells to which some animals are sensitive.

Just as canine mammals mark territory with their urine, hippopotamus, vicuña, rhinoceros, and probably a number of other species use their dung to mark territory. The hippopotamus's method is most interesting. As it is releasing its dung, the animal whirls its tail like a propeller; this movement breaks up the material and scatters it in all directions. Some of these dung-scattering animals have special glands at the anus to add a strong personal odor to the dung. The hippopotamus also uses its urine for marking.

Figure 9. Many mammals use scent glands to mark the boundaries of their territories and their mates. Some of these are located on the face, as in this black buck.[14]

Wild horses drop their dung around the outer limits of their territory, and members of the drove all contribute to this act. The dominant stallion, however, does his best to make the greatest contribution.

Small Mammals. We find quite elaborate systems of communication even among the smallest mammals. Some, like the porcupine and some rodents, appear to be quite silent. But many small animals' sounds are so high-pitched that human ears cannot detect them. This is true with mice and many other rodents; the distress calls of their young are much too high for us to hear. But a great many small mammal sounds are within our hearing range, and many of these have been studied.

Prairie dogs,[35] which are not dogs at all but ground squirrels, have been given their name because they utter a warning or danger signal like a dog's bark. Whenever a prairie dog barks in this way, it is really saying, "Get below ground!" All members of its group or family then head for their burrows or at least sit up and look all around them for the cause of the alarm. By changing its pitch a look-out animal can change the meaning of the bark to "all clear."

The yellow-bellied marmot,[69] another squirrellike rodent, uses odor, touch, and visible signals as well as sound. It has six different whistling calls with both long and short pauses between them. There is a quiet whistle that may simply keep members of the group in touch with one another. In another whistle the pauses gradually get shorter and shorter; this is for threatening and scolding. A third whistle with short pauses is used when a group of marmots are close to each other. All these calls are used for some kind of danger or to alert other marmots in the area. Marmots also have a barking whistle used while running, a single extra-loud whistle, and a scream. Visible signals are also used by marmots, and these are seen in their postures and movements.

Prairie dogs and marmots are not active at night, but many other rodents are. A number of these rodents use their vision much less than their other senses for receiving messages. They find it more useful at night to use their noses and their ears. The more an animal's nose and its center for smell in the brain are developed, the more active it is likely to be at night and the more it will use odors for communication. It is not surprising that mice have odors that are so personal they identify the sex, the species, and the individual. The males also produce in their urine a pheromone that conditions females for mating merely by its odor. A pheromone is a glandular secretion that stimulates the opposite sex. This pheromone is capable of making a female seem to be pregnant even if she is not—a false pregnancy.

The gerbil [76] is a desert rat that uses odor from a gland in the middle of its underside for several purposes. This odor also carries both an individual and a species label, and the gerbil uses its odor to mark its territory in the same way as a dog does with its urine.

Bats. There has been a great deal of publicity about the investigations into the echolocation ability of bats, but little has been said about their ordinary voice communication with each other. Many bats have large larynxes and masses of larynx muscles, and perhaps all use their voices for close communication. Although some of their sounds have been recorded, little is yet known of their meanings.

Sometimes the meaning of the sound is obvious because of what happens when it is uttered. For instance, the male white-lined bat [28] of Trinidad barks at every other male approaching his few square inches of roosting territory; this is a challenge. But as soon as a female comes near, he changes to a twittering song consisting of a mixture of chirps, buzzes, and pure notes that our ears cannot separate easily.

One of the most interesting things about this bat's voice signals is the fact that the frequencies of the notes are all multiples of each other. The male calls the female with a note of 6,000 cps. While hovering its sounds are 12,000 cps. It defies other males at 24,000 cps., and it hunts with 42,000 cps. Why all these signals are in frequencies so exactly spaced in multiples of 6,000 cps. is as mysterious as it is remarkable.

There is a very unusual fruit-eating bat called the hammerhead [55] which lives in Gabon. Some of the range of its voice is easily heard by human ears. The hammerhead is very large and ugly with a three-foot wingspan and a head like a moose; and it hangs in dense forest and gives off a loud, metallic honking from sixty to one hundred times a minute. This soon becomes a chorus as other members of the species within hearing range join in.

Such calling in bats is unusual and so is such great power, but the power is not surprising because the animal's larynx occupies 20 percent of its entire body volume. On a similar scale, a man would have a larynx almost half the size of his chest cavity. One can imagine that if anything like that were possible, the volume of noise it would make might shatter windows. The male hammerhead honks to attract females. Once a female approaches, the male stops honking and starts a repeated buzzing sound that is his way of singing a love serenade. But this bat's voice is not used just for mating calls. The bat will snarl quite viciously at any intruder—other than a female—to his roosting place.

Kangaroos. Marsupials as a whole do not have many sound signals, in spite of the fact that many of them have quite large larynxes. These large larynxes are of little use because marsupials have very weak larynx muscles or none at all. Some will grunt and scream if they are hurt, but that is probably the only sound they ever make. Monotremes (animals that lay eggs but suckle their young) seem to be completely silent. All marsupials have to use other means of communication.

Kangaroos signal alarm and danger by thumping their hind limbs on the ground. This thumping can be heard for some distance, and all within earshot will scatter. Body position gives important messages when used by kangaroos and wallabies. Their "alert" signal is quite unmistakable; they sit or even stand rigidly upright and still, their ears erect and turning in all directions as they explore the air for the direction of any sound, and their noses twitching as they test the air for odors. If danger becomes obvious, a big male will thump the ground, and the whole group will leave the spot with great leaps. When one kangaroo faces another and adopts an upright position, it is a challenge. If he crouches down again, the tension eases.

A rare nocturnal, insect-eating mammal called a solenodon or paradox [102] makes very unusual sound signals. Included in four kinds of voice sounds it makes are high-frequency clicks from 9,900 to 31,000 cps. in very short bursts of only thousandths of a second. Some of these sounds are what one might expect for sonar or echolocation like that used by bats, but we cannot be certain that the animal uses these sounds

Figure 10. This quietly feeding kangaroo family has been disturbed by a sound in the bush. The two leading males are testing the air with their noses and listening alertly. The movement of their ears in different directions can clearly be seen.

Figure 11. These tiger cubs are showing both fear and threat as the author enters their cage. The ears go down in fear, and the lips are drawn back in threat. The white identification spots on the backs of their ears can just be seen.

for this purpose. The paradox also has a highly developed olfactory (smell) center in its brain and glands for producing a strong musk odor, so it obviously has several ways of communicating with its fellows. Because this animal is now very rare, we may never learn why it uses high-frequency pulse signals.

Moles [105] appear to communicate with each other in their burrows with twittering sounds—which may also be "piped" by the sides of the burrow, like the digging sounds they hear when being hunted by larger animals. Their alarm signal is quite a squeal, in spite of their small size. They make the twittering sounds again when testing their surroundings, and there may be some echo-judgment in this. The moles may be testing the length of a burrow or the presence of any obstacles or other moles in it.

The male beaver has an unusual way of marking his territory with odor. He makes cakes of mud and fragments of wood all around the boundary of his home area and then scents them with a powerful gland secretion called *castoreum.* It is possible that the mud cakes keep their smell longer than the dog's urine.

Throughout the animal world are body-mark signals such as the white tip of a leopard's tail. The cheetah has this signal too; when it is raised high, it is clearly visible. The tips of a tiger's ears are also strikingly white, and a flick of these is quickly noticed by cubs. In fact, even when a tiger is so still that it is invisible to us, these marks will still stand out. Many other such markings are seen among mammals, such as the white-tipped tails of rabbits and a number of deer and antelopes. In fish these markings are used for recognition, for disguise, and for warning; and they are probably found more in fish than in all other animals put together.

We can only skim the surface of communication in mammals. To gather all that has been discovered into one volume would be impossible because there is not one species that does not have some kind of language adequate for all the kinds of communication needed in its particular way of life.

7. *Marine Mammals*

As water is a better conductor of sound than air, it is perhaps natural that marine mammals have learned to use their voices much more effectively than land animals. This ability has probably also been encouraged by the fact that visibility in water is less clear than on land. In the last forty years marine mammals have been very closely studied, and the information that has been obtained shows that they have amazing voice ranges and clever ways of communicating.

Dolphins. The dolphins have been the most rewarding animals used so far in these studies. They show all the emotions of excitement, fear, pleasure, inquisitiveness, welcome, teasing or humor, concern, and distress. They are also able to transmit and receive sounds over several miles, probably much greater distances than man-made underwater sound-detection instruments have been able to measure.

Dolphins have two separate ways of making sound, which are ordinarily used under different circumstances, but in courtship or play they may be used together. They consist of whistles, clicks, and buzzes used for echolocation, and of animal voice sounds. Almost all small-toothed whales—including dolphins—make whistles and chirps. Some also produce squeals.

Not only can dolphins make a wide range of notes, they can also use a wide range of volume, the difference between two sounds sometimes being as much as one hundred decibels. This is almost the limit of what a human ear can stand without discomfort. And the complicated way their sounds are put together varies according to whether the animals are alone or together in groups. One thing that never seems to vary is their distress signal—a repeated double whistle, the first getting louder and the second getting softer. Other dolphins will locate this sound, but the increase and de-

crease in volume makes it impossible for sharks to track it down.

When they hear this distress signal, all dolphins within hearing range become silent and search for the animal sending it. When they reach the distressed animal, they push it to the surface to breathe, all the while chattering to it in whistle tones. This combined effort will keep a sick dolphin breathing for days, even weeks, and even dolphins that are complete strangers to each other will cooperate to do this without ceasing until the sick or injured animal is well or it dies.

This response to distress is not something unique to dolphins. Many whales will also answer appeals for help—especially sperm whales. It has been noticed, however, that while bulls will help cows, cows will never help bulls except among humpback whales.[70] Cows will only help distressed calves. A film made by Jacques Cousteau on his research ship *Calypso* showed twenty-seven female sperm whales gathering from great distances to rescue an injured calf in answer to its distress signal. On land such a habit is less common, and perhaps the elephant is the only animal to help another of its kind.

Just as there is a definite distress signal like our SOS among dolphins, there is also a very recognizable sound of fear—a loud, sharp crack. Something like this has apparently been identified in other whales too. Each individual dolphin has a personal whistle pattern like a signature; other dolphins it knows recognize it by this whistle. Communication between dolphins is so good that they can even pass each other instructions or organize a herd to kill attacking sharks.

The killer whale,[86] which is actually a giant dolphin, also has a good vocabulary but with entirely different sounds. Instead of the whistlelike squeal used by other dolphins for communication, this animal has much harsher screams. Killer whales have larynxes but no vocal cords. They, nevertheless, have one of the most complex languages in the en-

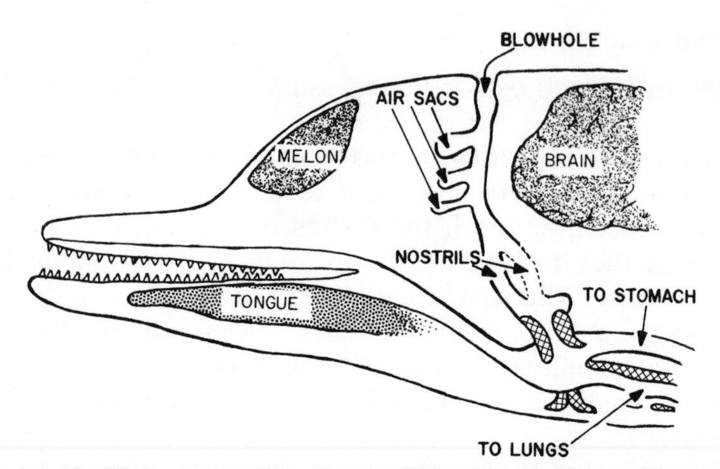

Figure 12. The two nostrils of the dolphin's ancestors have joined to form a single blowhole in the top of its head. Sound can be created by forcing air back and forth between the lungs and air sacs. The "melon" in the front of the head may be used to amplify the sound.

tire animal world, and hearing is their most important sense. They and the whales have taken full advantage of the fact that sound travels better, louder, and faster in water than in air.

The fact that sound is louder in water is obvious to anyone who lies in a bathtub of water and while keeping an ear under water knocks on the bottom of the tub. The sound will be quite loud, but it will barely be heard if the head is raised out of the water. Similarly, any sound made in the air is hardly heard by a person or animal underwater. If it is to be heard clearly, the sound must be amplified considerably or be high pitched. For this reason whistles have been used for thousands of years to attract whales near enough to be killed.

Very small underwater explosions can be picked up at a great distance. Because such noises are so greatly amplified, the Japanese used to be able to drive whales into bays by beating against the sides of their boats with wooden hammers. Anything falling into the water beside a whale creates enough noise to send the whale off in any direction at high speed.

Sound travels up to a mile a second in water, with slight variations for salinity, depth, pressure, and temperature. This is five times faster than in air, so a sound wave of any given frequency has to be five times longer than it would be in air. Water is therefore an ideal medium for sound communication, and other marine mammals—the seals and sea lions—have taken full advantage of this fact.

If dolphins have no vocal cords, how do they make their sounds? Some scientists believe that the sounds are made by forcing air from small sacs close to the blowhole (see Figure 12), and that small pieces of cartilage in the air passage play a part in their sound-making. Others think that the air vibrates in folds in the linings of the larynx and throat and that sounds are made when air is forced through the larynx, but our knowledge of this is still incomplete.

Sounds may pass through and resonate in a fatty, melon-like ball of tissue in front of the sacs because the animal's blowhole must be tightly closed while it is underwater. This tissue is connected to the brain through a nerve (trigeminal) that also serves the eye and other parts of the head. When sounds are made on the surface of the water, the dolphin's mouth moves quite distinctly, and some sounds do seem to come from the blowhole when it is open. Thus, there may be two pathways for the transmission of the dolphins' sounds, but this is still not fully understood.

All that we can be sure of is that the dolphin transmits sounds between 7 and 15,000 cps. with continuously changing pitch for conversation and between 20,000 and 170,000 cps. in short bursts of variable duration for its sonar (as some people prefer to call its echolocation).

There may also be some rapid passing back and forth of air between the lungs and the air sacs. This is done by frogs, so why not dolphins too? So far as other whales are concerned, it is not certain that their sounds are produced by a larynx, and it is not known whether one sex, or both sexes make the sounds, although most observations suggest that both do.

Hearing Without Ears. Although echolocation is not part of what we might call communication, the echolocation used by dolphins for detecting prey and obstacles does not appear to be received in a different way from the low-frequency sounds it uses for communication. Low-frequency sound travels farther in water than high-frequency sounds do, so the high frequencies are the ones used to detect prey or obstacles in the dark, and the low frequencies are used for conversation. When using its echolocation, the dolphin must be able to separate its own echoes from those of other dolphins in the vicinity, as well as ignore all the background noises that it will pick up. This whole hearing process is very complicated, and the separation process must all take place in the brain.

In spite of its great discriminating ability, the dolphin must hear in a different manner from land animals because its ears are tightly closed and the ear bones and the drum are quite loose. Sound evidently reaches the middle ear chamber through the body tissues and skull instead of through an outer ear chamber. Although the animal hears well in water, it might not hear so well in air. The effect is probably the same as in many fish, but we can not be sure of this because the oval window between the dolphin's middle and inner ear chambers provides a calculated amplification of thirty times.

The nerve from the dolphin's ear to its brain is unusually large, and the hearing centers in the brain are four times as large as the visual centers. This is the opposite of most land animals of similar intelligence. The dolphin's hearing system has actually reached a higher level than that of any other mammal, and this animal can measure distance and judge direction underwater with extreme accuracy.

Although the dolphin's method of hearing is somewhat mysterious, we can judge its effectiveness because, as was pointed out earlier, the development of a sense can be gauged by the size of the center in the brain that it serves. Even if there were no other evidence, the large size of the dolphin's hearing center tells us just how effective the dolphin's hearing must be in giving the animal a comprehensive picture of everything that goes on around it.

The larger whales hear in the same way as land mammals but with differently arranged ear structures, which are not quite the same as those of the dolphins just described. Some larger whales have a very narrow external ear channel, about 1½ millimeters in diameter, and they have an eardrum and a large ear bone; otherwise the middle and inner ear chambers are different from those of land mammals. The ear bone, which holds the drum, is only loosely attached to the rest of the skull by ligaments or connective tissue. There is no solid connection between the ear bones and any part of the skull, but it is this looseness that is thought to be responsible for the whale's ability to judge direction underwater. It is obvious that there are still a number of mysteries to be solved regarding the hearing of these marine mammals.

Whales. So many whales have been found to use voices for communication as well as pulsed sounds for echolocation that it seems certain that all whales do this. One whale that has been studied is the humpback whale,[70] which has probably the largest range of sounds and signals of all the large whale species. Although this animal appears to have no vocal cords, it must be able to use considerable air power in its larynx, for its whoops, whistles, grunts, roars, and bellows carry for great distances and the range of its tones is wide—especially for high frequencies used for communication within the herd. High-frequency sounds are used for this purpose because they do not travel as far in water as low-frequency sounds, which can be used to find the location of, and get information from, other whales many miles away.

Certainly it seems possible that the low-frequency calls of the finback whale[19] carry a tremendous distance. Roger S. Payne states that certain deep oceans will help the call to travel out an estimated 4,000 to 13,000 miles, the radius of an area of at least 50 million square miles. The purpose of such long-range communication is mysterious, but it seems to be a kind of "keeping-in-touch" with other members of the species that remain in the Antarctic home area while others roam the farthest oceans in pursuit of migrating food.

One of the more interesting things about the humpback whale is that it appears to sing real songs at tremendous volume. These songs have been recorded and analyzed and seem to consist of repeated phrases. They are not really songs, of course, but probably repeated messages, just as a ship's radio repeats a call until it is answered. Sometimes the sounds will last as long as thirty minutes—and never less than seven minutes, as has been recorded.

The baleen whale [82] uses lower and longer sounds than the toothed whales; the finback whale [19] mentioned earlier also makes low sounds, around 20 cps.; and the right whale [42] makes a belchlike sound lasting about one and a half seconds rather lower than 500 cps. All these whales seem to differ greatly in their signaling patterns. Although so many of their sounds have been recorded and their fre-

Figure 13. When on land the seal barks, but when underwater it uses many other sounds for testing its surroundings and communicating with its fellows.

quencies analyzed, very little is yet understood of the animals' actual language.

One mystery about the whale's voice communication is the fact that, for all its mighty size, its *external meatus* (the tube from the outside of its head to the eardrum) is only half a millimeter in diameter in some species, so it must have a middle ear that amplifies the sound greatly. Hearing cannot be through the body structure or bones, because all the blubber on the whale's body would act as a shield between the sound and the skeleton.

Seals. The voice range of the dolphin seems almost to be matched by the voice range of some seals and the sea lions.[116] Perhaps this is natural because of their similar habits. But some seals have poorly developed vocal cords, and others appear to have none, so how do they make their sounds? It seems that sounds are produced by varying the positions of a pair of cartilages called *arytenoids* so that the edges of these control the amount of air passing through.

Most seals, especially on land, signal with barks, which can vary greatly. Underwater there are also clicks, bangs, growls, and buzzes. The barks of seals are evidently aggressive and often a warning because when a large male barks, smaller males become more silent and move about less. When barking is of a fighting kind, it is usually toward males of about the same size. It is thought that this may take the place of real fighting in most cases.

In the sea these animals also use pulsing sounds for echolocation. But it was quite a long time before this was realized, even though it had been understood in the dolphins.

PART III

COMMUNICATION IN BIRDS

8. *Songs, Calls, and Display*

Birds use one of the most complicated communication systems in the whole animal world, and the variety of the sounds they make seems almost unlimited. They have calls and songs for many different purposes and situations. Being active mostly during the day, many of them also use visible signals, but odors appear to play little part in their lives. Small species, and those living among trees and shrubbery, use their voices more than most other birds, and this fact is also true in mammals. Those living where the visibility is good are the ones that use visible signals most.

The Syrinx. Birds do not use a larynx as mammals do. Although some birds are said to have a larynx, birds have no vocal cords, so sounds are not possible with their larynx. Apart from vultures, which appear to have no vocal organs at all, almost all birds have a syrinx at the bottom of the trachea where it forks to send a branch to each lung (see Figure 14). This is one of three basic forms of syrinx possessed by the majority of birds. These three kinds of syrinx have seven types of muscle structure to operate them, and this accounts for the very wide range of bird calls and songs we hear—from croaks, whips, bells, train whistles, clicks, and flutes to beautiful songs.

In the commonest kind of syrinx the trachea is surrounded by rings of cartilage, several of which fuse together right at the bottom to form a solid tube. But the first few rings of each bronchus are incomplete on their upper side, and the opening this makes is covered by a membrane which, with other membranes inside, vibrates to produce the bird's sounds. The range and quality of song is controlled by the number of muscles and their rapid change in tension. Most birds have a simple pair of *antagonistic* muscles (two muscles that act against each other), but singing birds have up to nine pairs.

Surrounding the syrinx is an air sac that is essential for the production of tones. It encloses air to keep a steady pressure around the syrinx, which otherwise would merely expand instead of remaining tense as air is expelled from the lungs into the syrinx. Without this sac efforts to get any real

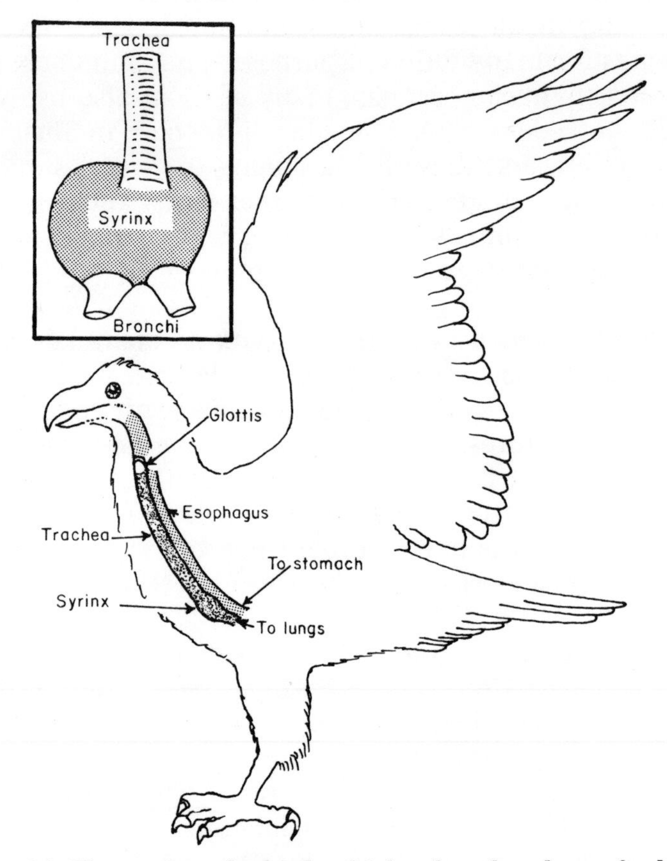

Figure 14. The *syrinx* of a bird, which takes the place of a larynx, is at the bottom of the trachea. The inset shows how the commonest form of syrinx is at the union of the trachea with the bronchi to the lungs.

Figure 15. Birds' ears are covered with fine feathers so their position and size is not seen. This South American rhea's ear is unusual in being clearly visible.

volume of sound would also put too much strain on the muscles of the syrinx.

Storks and many small South American birds have a syrinx that is confined to the lower part of the trachea and that is simpler than the one just described. A third form of syrinx is confined to the bronchi, and this is found in the oilbird [103] and the ani.[33] Some birds have developed devices for increasing the volume of sound they make. For instance, certain species of ducks have a large bony chamber at the base of the trachea that serves as an amplifier. In swans and cranes the trachea describes an "S" shape before dividing into the two bronchi, and the trachea again acts as a resonator or amplifier.

The Ears of Birds. Like those of amphibians and reptiles, birds' ears have no pinnas to direct sound into the outer ear chambers, as do mammals. The barn owl [112] is an exception, however, for it has flaps of skin that are not symmetrically placed, as ear chambers are. This makes it easier for the owl to judge the difference in the intensity of a sound as it reaches the two ears and to judge its direction more easily.

Birds' ears differ from those of other animals in other respects. The bird's eardrum, which is hidden behind a thin layer of fine feathers, is convex. Instead of three small ossicles (mallus, incus, and stapes) connecting the eardrum to the inner ear chamber, the bird has a single, rodlike *columella* like that of reptiles. This is efficient, however, because birds can hear notes that are outside our range and over distances much greater than we can hear.

The Voices of Birds. Some birds, such as pelicans, are more or less unable to make sounds. At the other extreme, there are some birds that call even before hatching from the egg. While still in the egg these birds hear their parents calling and learn to recognize and copy these calls. In fact, some unborn birds will even chirp if the temperature of the egg drops slightly—a warning to a parent bird to get back on the nest.

Certain species have several different song patterns; and many have different "dialects," due to groups of a species being separated by areas such as mountains or deserts, which are difficult to cross. This is also true of human language, for different dialects of English are spoken by people in Boston, Tennessee, and Texas. When birds live in dense vegetation, they must rely almost entirely on sound to keep in touch with each other. Their songs identify their species, mark their territorial limits, and accompany courtship and mating. Other calls are used as warnings, and these may vary according to the kind of predatory animal that is seen and sometimes will even tell the location of the predator.

Birds have different sounds for conversation, for telling others of a food discovery, for anger or scolding, for attack, for victory, and for distress. Other calls carry information between pairs, between parent and young, between only the young, and between members of a flock. The young also have definite food-begging and distress calls and sounds for contentment.

Most true bird songs are sung by the males, although in some tropical areas females sing too. In species where the female also sings, each pair has a personal duet pattern that does not depart from the main song of the species.

Learning a Song. It has already been suggested that the ability to learn to speak (or to make whatever sounds an animal species needs to make) may be inherited, but a particular language (or song) must be learned from other members of the species. This has been proved several times in birds by making them deaf when young. These birds never get beyond the first simple stages of song development. If they are kept away from other birds as soon as they are hatched but without being deafened, young birds develop their voices but they never sing the true song of their species.

If several of these isolated birds are later put together, they will compose their own song, which is quite complicated but not the same as the song of the rest of their species. However, this is certainly not true of all species. It seems that doves and chickens that are deafened just after birth can still use the natural species sounds, but these birds may learn the sounds while still in the egg. In any case these sounds are simpler than a song.

Further proof of the need to learn a song has come from work in the USSR, where the eggs of one species of bird were placed in the nests of other species. When these eggs hatched, the young birds grew up singing the songs and making the calls of the species that reared them (their foster parents). This happened even when they were within hearing distance of calls of their own species. However, the birds also used certain calls that are evidently inborn for their own species; this may be because of the inherited form of the syrinx and its muscles. We cannot be sure of the reason.

If birds are deafened after they have learned their true song, many continue to sing this song quite accurately. If experiments like those carried out in the USSR were conducted on certain birds, they would be quite unreliable. For instance, the Australian lyrebird [75] frequently mimics almost every sound it hears, whether it is another bird's call, a human sound, or even a train whistle. It is doubtful that these birds understand what they are mimicking, nor is it clear why they do this. Other birds also do this. The cuckoo is exceptional because it does not mimic.

Territory and Mating Songs. When male singing birds mark out their territory with their song, many fly from point to point around their boundary, giving their song at each point to advertise their limits and to warn other males to keep out. The denser the vegetation, the louder they sing. Not only does this keep out other males, it also attracts females. There is something in each species song that pleases the females of that species. In fact, an individual male may add something special to his version of the song, because it is known that a female will sometimes refuse to mate with any other male while she can still hear the song of the particular male she favors.

Not all birds go from point to point around their territory. Some warblers sing from a single tree in the center of it and a female warbler "homes in" on the male's song like an airplane homing into an airport in fog on a radar beam. Only a song that is exactly right for the species will attract females of that species. Any male that is not note-perfect fails to breed. If this were not so, there would be confusion in the world of birds. The female must also make the correct responses for successful breeding. In some species, when the male sings the proper song, the entire absence of any response by the female may be the correct signal for him to continue courting her. Both he and she must make the right signals and responses.

Some bird species have many different song patterns. The Carolina wren,[110] for instance, has twenty-two. The song sparrow [74] has sixteen with seventy-five or more kinds of syllables. Many birds vary their song patterns by dividing them up into different combinations of notes in different orders, but the combinations are always kept within a certain range. Some birds even have the ability to add something to their songs, but again the main species theme is always there.

Calls With and Without Song. Certain birds make no sounds that we would call song, but they have calls with which they keep in touch with each other. These birds do not usually mark out territory, so they may have less need for song. Just keeping in touch within the group will bring them together for breeding anyway, and calls can be used for that.

Even birds that have territorial and courtship songs also have certain calls of this kind, and their alarm call is usually the one most easily recognized. But their warning signal for invading members of their own species is quite different from the one they give when humans or predators come into view. There are, in fact, many versions of alarm calls; some of them not only alert members of their own species but also alert other birds and animals, which even seem to understand the differences in calls for different kinds of danger. Warning calls can also be divided into those for possible danger and those for definite danger.

Indian mynas [3] have a special danger signal for humans and birds of prey which sends all other birds diving for safety as well, often giving their own alarm calls as they go. The chickadee sounds a different alarm call when danger is overhead than when it is on the ground. Crows have five different danger calls, and jackdaws even have one that brings help from other jackdaws, but this would probably also be considered a distress call.

These alerting calls are always the result of *seeing* an enemy. This fact has been proved in the laboratory by electrically stimulating the vision centers in a bird's brain. This stimulation automatically produces scolding and alarm calls and results from the fact that there are connections in the brain between hearing and vision centers.

Many birds vary the pitch of their danger calls to confuse hawks and other predators, and there is a reason why this is successful. The easiest sound to locate is one that is steady, like a "clucking" or "chirping," because these sounds can be timed by the two ears. But when calls are varied by fading or by gradually amplifying them in such a way that they seem to be moving farther away and then coming closer, they are much more difficult to trace.

Identification calls keep a flock together or bring it together, and some birds use a call of this kind when one of them finds a supply of food. Identification is very important between parents and their young as well as between the parents themselves. Thus, these calls are often quite individual, although to our ears the differences may not be apparent.

Penguin calls to their young all sound alike to us—but not to the chicks. Within a colony of many thousands a parent's call will bring only her chick to her, so penguin voices must be as personal to penguins as ours are to us. There are some birds, however, in which individual calls are so different in tone that they can be identified by man; but even in these the species sounds are always present.

It has already been mentioned that certain birds call before hatching from the egg. Sometimes this goes on for hours or even days before final hatching. Air can penetrate an eggshell, and the chick uses air in the space between the inner and outer shell membranes to make the sounds with its syrinx. It responds to outside sounds, to the mother's calls, and to other embryos calling. This is thought to be a provision for making certain the eggs all hatch at about the same time because those in contact with each other do hatch at the same time but more distant ones do not. The hatching of those not in contact may be spread out over a twenty-four-hour period.

The rhea, an ostrichlike, South American bird, has very successful simultaneous hatchings in this way. It probably needs this ability more than many other birds because a number of female rheas visit a male and leave eggs with him to incubate and hatch. By the time he has all he can manage, there may be fifty or sixty eggs. The embryos communicate with each other, however, and although there may be quite some time between the laying of the first and last eggs, all hatch at about the same time.

Some of the bird sounds most pleasing to our ears are quiet conversations that go on between a pair in some species. These conversations appear to be quite personal. Perhaps the most familiar of these is that of the duck; when a female has young around her, she hardly seems to cease her chatter. The Australian magpie [46] converses so tunefully in this way that one cannot help but get pleasure from listening to it. At times the conversation is almost like a quiet little duet.

Many Australian birds talk among themselves all the time because Australia has so many parrots, cockatoos, and budgerigars. These birds also have their calls, which may vary with the seasons, but the chatter continues all the time. The Eastern rosella [95] has about twenty-four different calls for warning, distress, aggression, location, and victory in a fight.

Most birds seem to have more danger calls in autumn, when there are more predators about, and more aggression calls in spring, because of mating rivalry. Then as the juveniles and other groups gather together before the winter, aggression calls once again become more frequent.

The more we listen to birds, the more certain we become that most people miss the meaning of their language. Probably all birds have wider vocabularies than anyone realizes. Because the canary is a captive bird, we recognize many of its messages; it has ten or more kinds of sounds for courtship alone, as well as calls for attack, alarm, agitation, anxiety, and surprise. Chickadees [91] have sixteen songs and calls with obvious meanings, and even the humble domestic chicken, which seems so stupid to most of us, has more than twenty. All birds express in one way or another alarm, hunger, identity, finding food, territorial aggression, threat, attack, flight, courtship, and mating.

The Indian myna[3] (one species of which is called the noisy miner in Australia) is another constant chatterer. Some of its calls we can recognize easily. One of these is uttered by the bird when it finds food, to tell a widely scattered group about it. Another is a scolding cry to drive away competitors—larger birds, foxes, and cats—and to warn of snakes. The last call is also used to warn others of another group that have come too close. As explained before, this bird has a separate danger cry against falcons and humans. Another call is a kind of inquiry when danger appears to be past: "Is it okay now?"

Bird Displays and Signals. Bird display and bird song cannot be separated because in some species the two are so intimately linked that they occur together, reminding us of very similar behavior seen in some mammals. A simple example is the cat, which arches its back and hisses at the same time. This is display and voice being used together. We even see the same thing in young boys who put up their fists and say at the same time, "Come and get me." Yet in describing display we must concentrate on display by itself to some extent, for certain birds display without making sounds.

POSTURES. Many birds signal their feelings or their intentions clearly with the position of the body, tail, wings, bill, head, crest, and legs; frequently combining these with repeated movements, calls, bill-clapping, or quiet sounds. Those birds in which the male performs aerobatics to impress a female often sing at the same time, but not all do. Sometimes these acts do not carry a message at all. They just identify a bird or its species like a signature.

In any species in which the male and female are similarly colored, they may identify their sex only by posture or movement, but when a movement carries a message, it is clear to every member of the species. For instance, it has been noticed in Wales that the sandpiper raises its right wing to a vertical position as a warning, and it would be difficult for any other bird not to see this.

Display is an important part of courtship. While display may seem to us to be something that should be private between a pair of birds, some species congregate in groups on special display or "strutting" grounds for their courtship displays. Often these groups include considerable numbers of both sexes, the females watching the males strut around, spread their feathers and tails, and perhaps make throat sounds too. But not all strutting is done in groups. The pigeon, the ostrich, the peacock, the lyrebird, and many others require no audience at all, although if there is one, this does not affect the behavior of some of them.

Individual displays—that is, those in groups that do not congregate—include postures, movements, and color. These will often but not always go on with song or quiet chatter. Male birds of paradise have vivid displays. They perch, fly, or even hang upside down, spreading their tails, wings, and special feathers or plumes that show colors, shapes, and attitudes while singing or calling their courtship tunes.

Once birds have paired up, the displays change in many species. Greeting (recognition), especially at the nest for identifying each other, and the offering of gifts may take the place of the previous courtship displays. When the two sexes have the same coloring, posture can be most important for recognition. A good example of this is seen in the frigate bird,[43] where the male points his bill up in greeting and the female points hers down. The male penguin may stretch up his neck in greeting, while the female will bow hers.

Postures also convey other information. The black cormorant [92] crouches horizontally with its crest down when at the nest, but when it is away from the nest, its body is upright and its crest is raised. The crest goes down at once, however, to signal danger. It is natural that birds without voices or with syrinxes that hardly function should use visible signals, and this extends to basic things as well as to warnings and greetings. When the voiceless young cor-

Figure 16. The Australian wedge-tail eagle[114] very seldom shows a crest, but when angry or threatening another creature it raises a crest quite clearly, as shown here.

Figure 17. The male gannet points his bill to the sky to tell his mate he is about to take flight.

morants are begging, they close their bills for food and open them for water.

THE DRUMMERS. Woodpeckers can communicate in a kind of Morse code by drumming with their beaks on tree trunks; this has nothing to do with their pecking to get at insects. Woodpeckers may make only a single knock or several lasting a few seconds, but these knocks seem to carry a message to other woodpeckers out of sight. Male grouse make drumming noises too, but in their case it is done with their wings for courtship or challenging other males. This bird also makes posture signals by expanding its body and raising its feathers.

OTHER SIGNALS. Most attention has been paid in the past to the display and behavior seen in courtship and raising the young, and our knowledge of other visible signals made by

Figure 18. The male giant bustard [87] raises its tail and inflates its throat sac or pouch (as shown here) when threatening a rival.

birds and their meanings is not yet so comprehensive as it might be. There are probably nearly forty different voice and visible signals in the language of some bird species, and this is many more than would be required just for bare survival in courtship, alarm, and defense. Some signals used in disputes over territory are not displayed by males alone; in some species both sexes display together.

Among the special display features commonly seen are the inflatable pouches of the male frigate bird [43] and the giant bustard,[87] and crests that are invisible until raised in anger or display of the cockatoo.[56] Such movable crests are used for signaling. There are also crests that are always in a raised position and combs like those of the domestic chicken that are similar to the helmets of the hornbill or the cassowary and the shield of the marsh hen. These are only a way of identifying the species or of separating the sexes.

Some birds open their wings to reveal different colors or special markings when they are about to take flight. It seems that this movement stimulates a whole flock to take off at

Figure 19. The Nankeen night heron [84] suddenly raises its ruff or neck frill and crest as a threat to a rival or intruder.

Figure 20. This young Nankeen night heron [84] is reacting to another's threat by raising its ruff and rearing back in what is called a "fear-threat" signal like that of the young tigers in Figure 11.

the same time. Whooper swans [34] bob their heads up and down for the same purpose.

All animals have language of some kind or another and use it effectively.

PART IV

COMMUNICATION IN FISH, AMPHIBIANS, AND REPTILES

9. *Sound and Other Signals in Fish*

Anatomical Sounds. Fish have no larynxes, so all sounds must be made in some other way. Many fish use the gas bladder by vibrating various formations of drum muscles around it. In some, strands of connective tissue stretched inside the bladder are used to vary its tension and its resonance. Squirrel fish [52] are typical of fish in which the bladder is connected to the inner ear. In the carps [38] minute bones form a sound-conducting bridge between the gas bladder and the inner ears, the bladder again acting as a resonator that is apparently involved in both the transmission and the reception of sound. This coupling of both ears to the gas bladder produces the effect of having only one large ear, and while this may amplify sound, it also prevents fish from being able to detect the direction of the sounds they receive. Instead, fish detect the direction of the sound through the lateral-line system.

Many species of fish produce croaking or moaning sounds with specially modified fins, with their jaws, with their throat teeth, and with other parts of the skeleton. The sound is produced when two parts of the skeleton are pressed firmly together and then set in motion. In an Indian catfish [50] and in an African air-breathing catfish,[30] sound is produced by the movement of a pectoral spine or by rubbing together pharyngeal teeth on the floor and roof of the mouth or by both methods.

When gas bladders have no vibrating muscles, they are often arranged to resonate the grinding of these pharyngeal teeth. Such sounds are used by the fish for protection or for

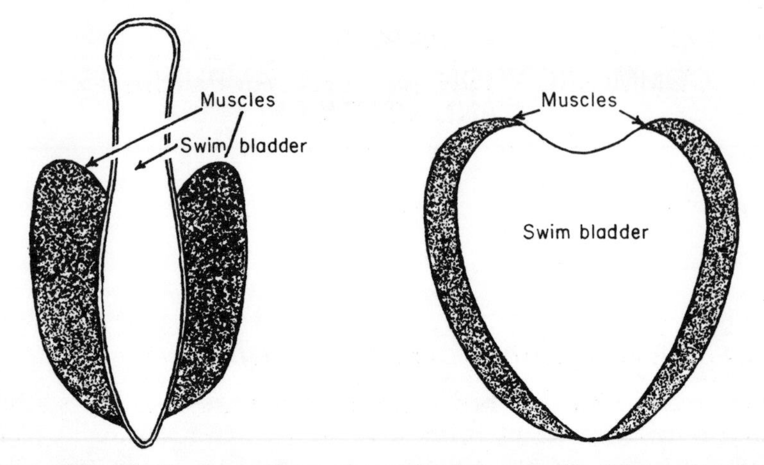

Figure 21. Vertical sections through fish gas bladders showing how muscles are located in some of them for producing tension and sounds.

shoaling together during the breeding season. The range within which all species of fish make sounds appears to be between 20 and 11,000 cps., but some have a wider range than others, and the range will even vary in different species within a family. While minnows can make sounds up to 11,000 cps., one finds a narrower range in most other fish, often only between 100 and 600 cps.

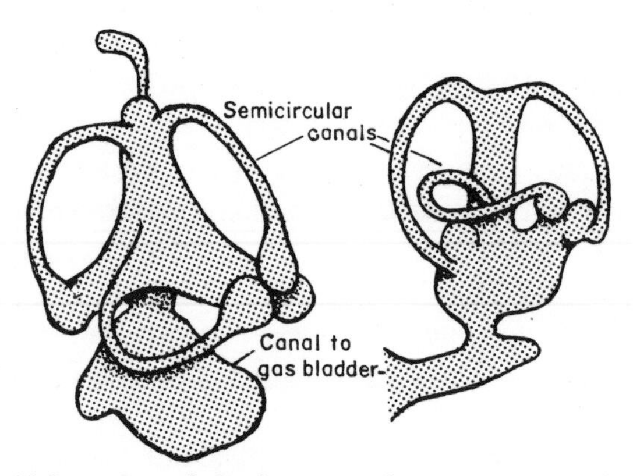

Figure 22. Although a fish does not have outer and middle ear chambers, it has a very sensitive inner ear with three semicircular canals controlling the fish's balance.

Fish have neither a true eardrum nor middle-ear ossicles like mammals, but their inner ear is well formed. It is frequently embedded in the bones of the head, and it was once thought that this organ was involved only with balance, but it is now well established that fish hear with it too.

In some bony fish the two internal ears are connected by a transverse canal, by ducts, or by a series of small bones (*Weberian ossicles*) to the gas bladder. This occurs in perhaps up to five thousand species of fish that have a gas bladder. Perhaps these ossicles are like those in the middle ears of mammals because they transmit movements of the gas-bladder wall to the inner ears, using the bladder as an eardrum.

This means that the hollow chamber in fish has two uses: both the reception and the transmission of sound, but without interfering with the control by the ears of other functions, as shown by the fact that when the Weberian ossicles are removed in carp,[37] the fish's ability to hear high-frequency

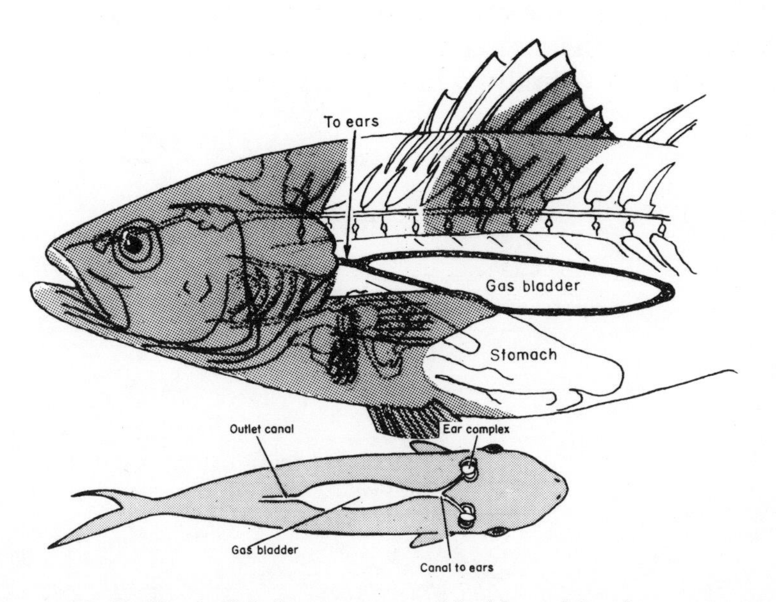

Figure 23. In those fish having a gas bladder, this often acts as a kind of eardrum, and tubes from this to the inner ears (below) convey its vibrations to them.

sounds and its ability to be aware of slow pressure changes in the surrounding water are both reduced.

Herrings [31] have tubes that reach from the gas bladder to the inner ears, and it has been shown that this is responsible for the herrings' keen hearing. Herrings are more sensitive to slight sounds, and they have a greater range of hearing than other fish. This may be unexpected in fish that are silent themselves, but herrings need this sensitivity because they are among the most hunted fish.

The function of the inner ears as gyroscopic organs of balance in fish appears to be related to sensitive cells in the lateral-line system, and some experts think that ears evolved from the lateral line. There is certainly evidence for this. Not all sounds received by fish reach the brain by way of the ears. In many fish experimented with, sounds reached the brain through the lateral-line system when the frequencies were low and through the ears when the frequencies were high.

Any center in the brain of an animal that is not of great importance either remains undeveloped or it degenerates, so the size of a brain center is a reliable indicator of the

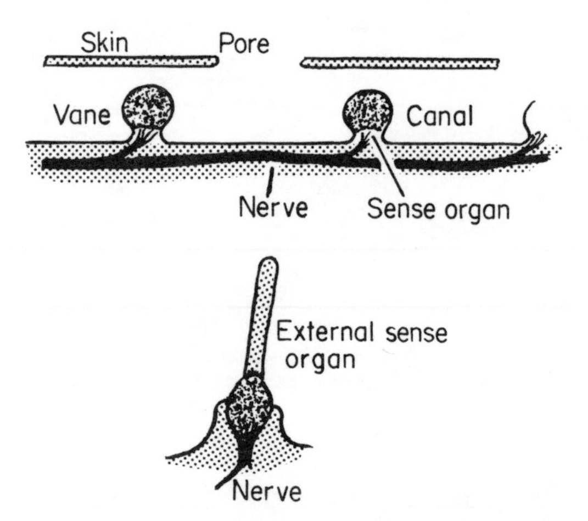

Figure 24. A diagram of the "pore canal" system, which may be a fish's receiving organ for low frequency sounds.

extent to which a sense is used. The size and development of the hearing centers in the brain of fish prove that fish can hear a wide range of sounds, so they may be much more sensitive than we imagine.

BONY FISH. The sounds made by some species of fish are used mostly for keeping or bringing a shoal (school) together. They are also used as recognition signals and for attracting the sexes to each other as well as warnings to rivals or enemies. A few of the better-known fish that use sound signals extensively are the haddock [44, 72] (which grunts and makes rasping noises for defense and aggression), the anchovy,[41] the Jewfish,[81] the weakfish,[36] the jackfish,[27] the pike-perch,[65] the ninespine stickleback,[98] and the grunt,[48] as well as salmon, bass, catfish, cod, mackerel, and many other familiar species. One might say that all fish the average fisherman is likely to catch make sounds of one kind or another.

The male of the northern midshipman [97] is even called the singing fish because if it is guarding a nest, it gives out a hum when alarmed. The fish hums by vibrating its gas bladder. This fish also carries light signals right down the length of its body. The sea robin [111] and other members of its family group are really gurnards, a name that is derived from *grogner* (meaning "to grunt"), and which was given them by the French. These fish make sounds that vary from grunting and snoring to crooning with their gas bladders, vibrated by sets of muscles. The sea bass, on the other hand, makes its sounds by beating its gill covers against the sides of its head.

Drum [100] make so much noise, especially at spawning time, that it can be heard in a boat. They, too, do this by vibrating the gas bladder. Croakers [77] are another group that drum like this. Another fish, which is also called a drum,[96] is one of the noisiest species known, and in the breeding season both males and females call. Although their calls are loud, they are not as clear as those of the other kind of drum.[100]

All kinds of strange noises come from fish when they are

disturbed. During their breeding season toadfish [85] toot like little ships, and the males grunt while guarding their eggs. Marine catfish bark like seals (but quietly), and squirrel fish [53] make chattering noises.

The purpose of all these noises is not known, but they do seem to include signals for members of each species to keep in touch with one another or to alert them to danger, because the sounds become much louder from some fish when they are caught on hooks. Some fish also make loud noises when driving away intruders, and experts who have studied these fish claim that the sounds are often so personal that individual fish can be recognized just as individual people can.

It would be impossible to name all the fish that use sound to communicate because they are found from the surface to the darkest depths of the oceans. One can understand that sounds made in the dark zones would be for species identification, keeping in contact, and for sex identification, but it seems to be just the same in fish that stay in the bright areas at the surface.

One gains the impression that fish have control over the muscles that vibrate the gas bladder, and fish that use this method of making sounds in the darker depths also have very well-developed hearing centers in the brain. This tends to confirm that they use sound for communication. Even fish with less well-developed hearing centers can pick up considerable information in the form of vibration through the lateral-line system. Certainly sound signals are used as much as, if not more than, visible signals, except perhaps in the well-lighted shallow inland waters; but even there a group of fish can be quite noisy at times.

An unusual method of drumming is seen in the triggerfish,[21] including the Hawaiian black triggerfish.[73] The gas bladder is closely connected to skin, forming a tight membrane behind the gill opening and above the base of the pectoral fin. The fin is beaten rapidly against the surface membrane to produce the sounds.

Russian scientists have searched for explanations of fish

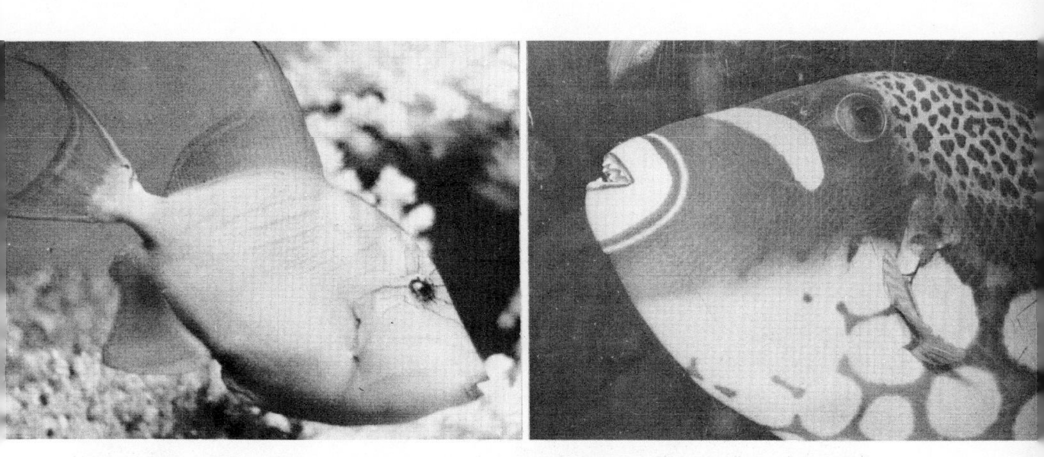

Figure 25. The queen triggerfish [22] and the clown triggerfish [20] have their air bladders connected to a membrane behind the gill opening and above the base of the pectoral fin, which is beaten rapidly against this membrane to produce sounds.

sound signals, and they have identified at least three kinds of common signal. They have also measured how far these signals will travel and still produce a response in other fish. The scientists found that sounds produced when danger was close or when feeding attracted other fish from as far away as one and a half miles, while sounds produced for spawning brought in others from even greater distances.

One thing that the Russian experiments showed, and which had already been revealed when experimenting with dolphins and whales, is that the distance from which a fish will respond to signals by other fish depends on the amount of background noise through which the signals must penetrate. The more noise there is in the area, the shorter the distance a signal will be heard. Sounds for offense, defense, and intimidation do not have to travel so far, so background noise is not important when they are used.

Blind fish, of which there are many species living in caves or in the ever-black ocean depths, must obviously have some special way of communicating either through pulsing electrical fields or through sound. Such sounds are now being identified. A blind cave fish [10] in Kentucky can make sounds with its gas bladder, which it can also use for echo-

Figure 26. Blind cave fish [13] sometimes have no trace of an eye and other times just have a small pigment spot where an eye would ordinarily be. Both are shown here. This fish's communication and echolocation are through its gas bladder.

location, because it has particularly well-developed lateral-line organs. Another fish [104] that behaves in a similar way seems to use sound for all its communication and echolocation needs.

SHARKS. The hearing of sharks has been carefully studied because of the need to understand whether it has anything to do with their attacks on man, but little has been said of any sound signals sharks may make themselves. Skin divers who have had the courage to swim with the forty-foot whale shark say that it makes a constant croaking noise. This could be a kind of sonar like that of the dolphin or the killer whale, but as this shark eats only plankton, it may not be for the detection of prey, although a large body of plankton will reflect an echo of sorts. Such low-frequency sounds travel a great distance in water, so they may, on the other hand, be for keeping individual whale sharks in touch with each other.

So far there are no records that suggest sharks as a whole are able to make sounds or even have really useful hearing

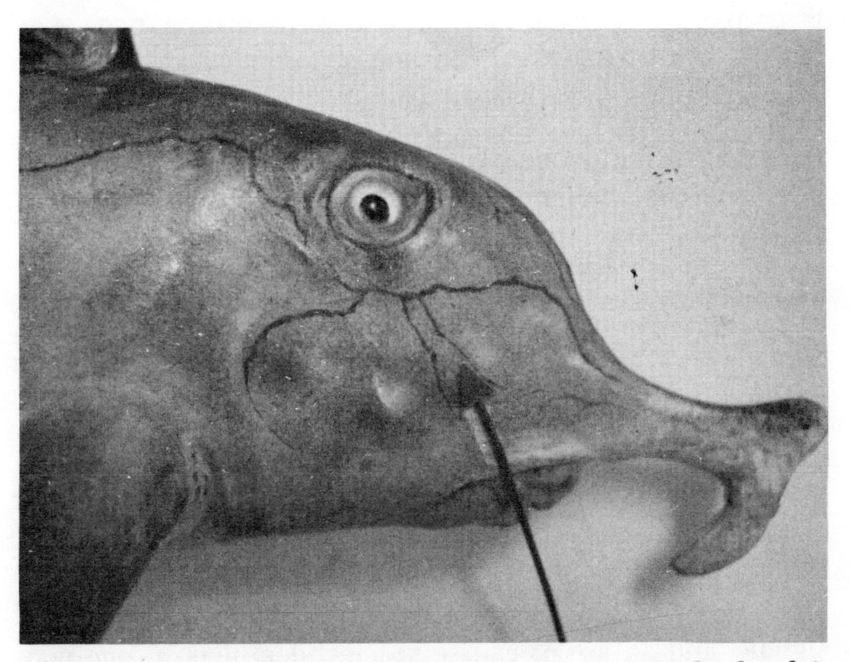

Figure 27. The branching of the lateral line over the head is clearly visible in this bottlenose chimaera shark.[25] This shark is said to be sensitive to certain low-frequency sounds, but it is also highly sensitive to odors, electrical currents, and water pressure.

centers, even though they can be attracted by low-frequency sounds picked up through their lateral-line systems. Sometimes sharks take in air at the surface, and the expulsion of this air makes a loud belching sound, but this has no communication value.

Visible Signals Made by Fish. It is possible that as many species of fish use visible signals as use sound for messages. Various species have postures that indicate danger, defense, defeat, courtship, spawning, and calling together their young. In addition many are able to vary their colors, and many others use luminous organs to flash messages or identification patterns.

COLORS. Color change can be slow, as it often is when a fish is preparing for courtship, or it can be instantaneous, as when a fish is frightened or when a male meets a female or a rival. Fright may make a fish turn pale, but anger and rivalry will flush its colors to their limit. All such changes are the result of either hormone influence or nervous stimulation, and the change takes place in tiny cells in the skin known as *chromatophores*.

The chromatophores expand with pigment granules of orange-red, yellow, and a kind of black-brown called *melanin* in combinations that will produce almost any hue. White and silver will combine with these granules in the form of small reflecting cells called *iridocytes*. When a chromatophore contracts, the pigment is drawn tightly into the center of the cell, but when it expands, tiny processes extend out from the cell and the pigment expands into these to cover the other tissues (see Figure 28).

Most fish coloring is probably used for camouflage. Some color markings are for sex identification, some are breeding signals, and all are of course species identifiers in one way or another. There are a few fish that change color when it becomes dark, but this cannot be any kind of signal; perhaps

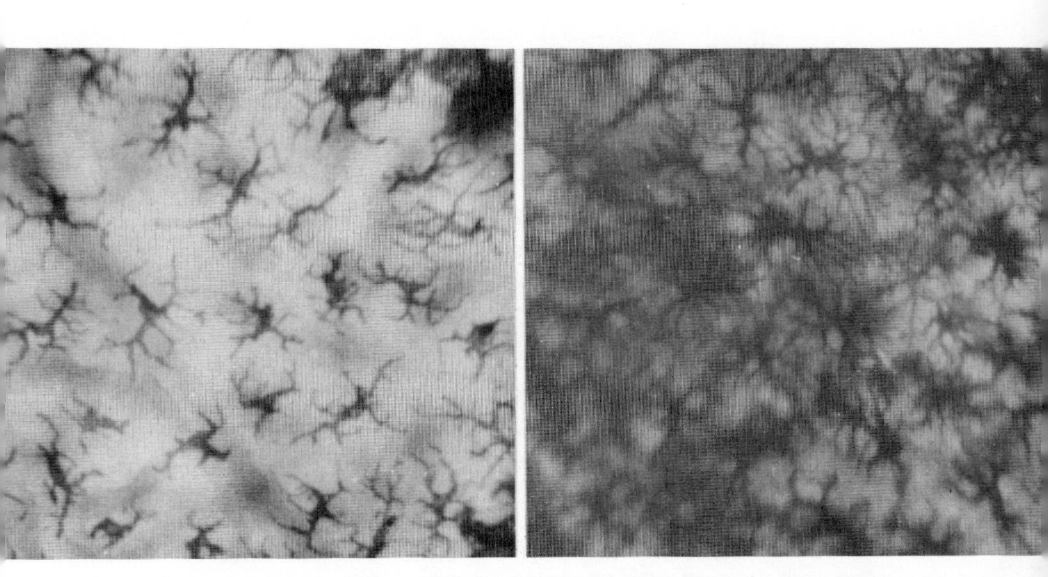

Figure 28. Chromatophore pigment cells in the skin almost contracted (left) and expanded to give more color (right). These are controlled by hormones and by nerve stimulation.

it is no more than chromatophores returning to a position of rest.

Besides color change there are many ways in which fins—especially the dorsal fin—can be held to convey a message. Many fish will open their gill covers wide as a message inviting small "cleaner" fish to search them for parasites and remnants of food. There are also signals for the cleaner to enter the fish's mouth and different signals for it to leave.

MESSAGES OF LIVING LIGHT. Although many creatures that are active at night or in the dark depths of the ocean may use sound for recognizing others of their own species or to convey messages, those that retain vision as their most important sense may still use visible signals. Many fish produce their own light, either in cells or in body organs. The light can be switched on and off at will or flashed in a message pattern. The advantages light signals may have over sound or odor are not entirely clear, except that in deep-sea fish light makes possible a much wider range of signal patterns

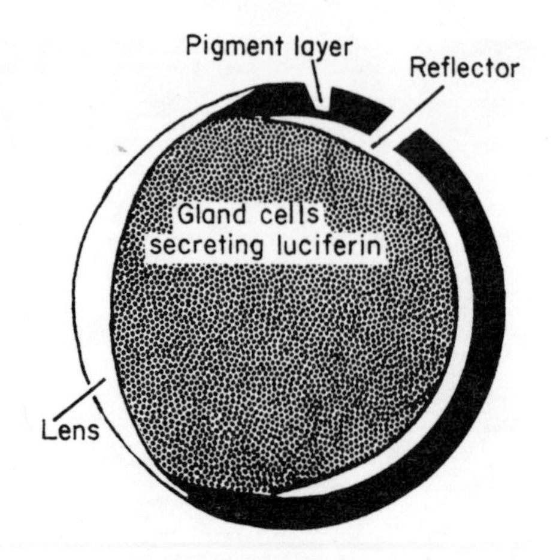

Figure 29. A photophore or light organ found in fish and insects consists of a gland which secretes *luciferin,* a chemical that lights up when oxygen reaches it. A reflector and lens direct the light and concentrate it like a miniature projector.

than is possible by any other means. The number of possible light patterns is limitless.

Many of the fish living in the deeper oceans that have light organs have developed the sensitivity of their eyes to an extreme degree. They live in perpetual darkness, so they have not had to concern themselves with apparatus to protect their eyes during the day. Instead, they have been able to concentrate entirely on developing this sensitivity and maximum light collection. A number of crustaceans also have light organs, but it is thought that in these animals the lights are used only to attract the sexes to each other or to bring large numbers of a species together, perhaps for mass mating, but this also reveals them to predatory fish.

In some layers of the oceans up to 95 percent of all species have luminous organs. No fish with luminous organs, however, are known below a depth of 16,000 feet, where eyes have usually been sacrificed by most of the inhabitants. In fish that produce light, the photophores are usually under their control, permitting the fish to douse its lights when danger approaches, or to flash them for messages.

This living light is produced by chemical action in which a body substance called *luciferin* is combined with oxygen and an enzyme called *luciferase.* The luciferin is secreted by a gland, and the oxygen is brought to this gland by a close network of blood vessels. When the flow of blood in these blood vessels is reduced by tiny muscles, the oxygen supply is cut off and the light is extinguished.

While the manufacture of this light is always through luciferin and oxygen, there are slight variations of the luciferin in different species, and the luciferin, luciferase, and oxygen are frequently combined by different methods. This is especially true in crustaceans. The gland that manufactures the light not only secretes the luciferan for this purpose, but it also acts as a projector for its own light, using a mirror and lens to do this (see Figure 29).

There are two other ways in which fish produce light. In one, certain groups of gland cells in the skin secrete a kind of slime, and as this meets the water, it takes up oxygen and glows. In the other form, colonies of phosphorescent bacteria live in pits in the skin, and these bacteria glow on contact with water or when oxygen reaches them through minute blood vessels. These light organs are always arranged so they will never throw light directly into the eyes of the animal that carries them.

One of the most important qualities of this physiological light is its burning efficiency. At least 90 percent of the light becomes cold light, and almost none of it forms heat. This is much more efficient than any known man-made form of artificial light, many of which convert no more than 5 percent of their total energy into light.

The lights vary in hue as well as in the patterns they form on the body, and the hue always coincides with the wavelength of light to which the eyes of the species are most sensitive. So apart from those that use light as bait to attract prey, the purpose of the light must be mainly for recognizing their own species. The hue may also separate one species from another that has similar light patterns.

There are ways in which a fish can be made to lose control over its lights. When water is charged with a powerful electric current or when ammonia is added to the water, the lights are brought on at once. It may be a signal of distress like that in fish that change their color when they receive an electric shock, but the importance of this reaction is obvious when there is a possibility of electric fish being around. Another thing that will bring on the light in all of a fish's photophores is the injection of adrenaline. This shows that photophores are affected by hormones as well as by electric currents or chemicals in the water.

These lights may have several purposes. They enable fish to recognize their own species, bring the sexes together, or keep a group together either by the pattern of the lights on the body, by their flashing sequences, by the color of the light, or by a combination of any of these. In some species, as in some fireflies, only the male flashes signals; this can only be for sexual and species recognition. In other species it may also be for attracting predators away from females that are important to the survival of the species. There is nothing unusual in this; in many animal groups the male sacrifices himself so a breeding female may survive. Even man does this.

Although it might appear that these luminous organs will attract predators, this disadvantage is outweighed by a less obvious advantage. In regions without light all creatures will gulp down any others. Since it is ordinarily one of nature's laws that a species should never kill its own members, recognition marks are essential, and these lights are the species' identification. It is true that some fish eat their own young, but it is possible in these circumstances that they do not recognize them until their bodies have assumed adult markings, color, or form.

Light organs are found on certain deep-sea sharks as well as on thousands of species of bony fish. Although in sharks the light is also produced and projected by a gland behind a lens, there is no reflector, so the light is not usually so bright as in bony fish. Often the lights are very small but in such large numbers that the shark is given more of a glowing outline than a pattern of small, single lights. This again gives us the idea that such lights are mainly for recognizing members of the same species and above all for avoiding attack by one shark on another of its own species when an encounter takes place in the dark.

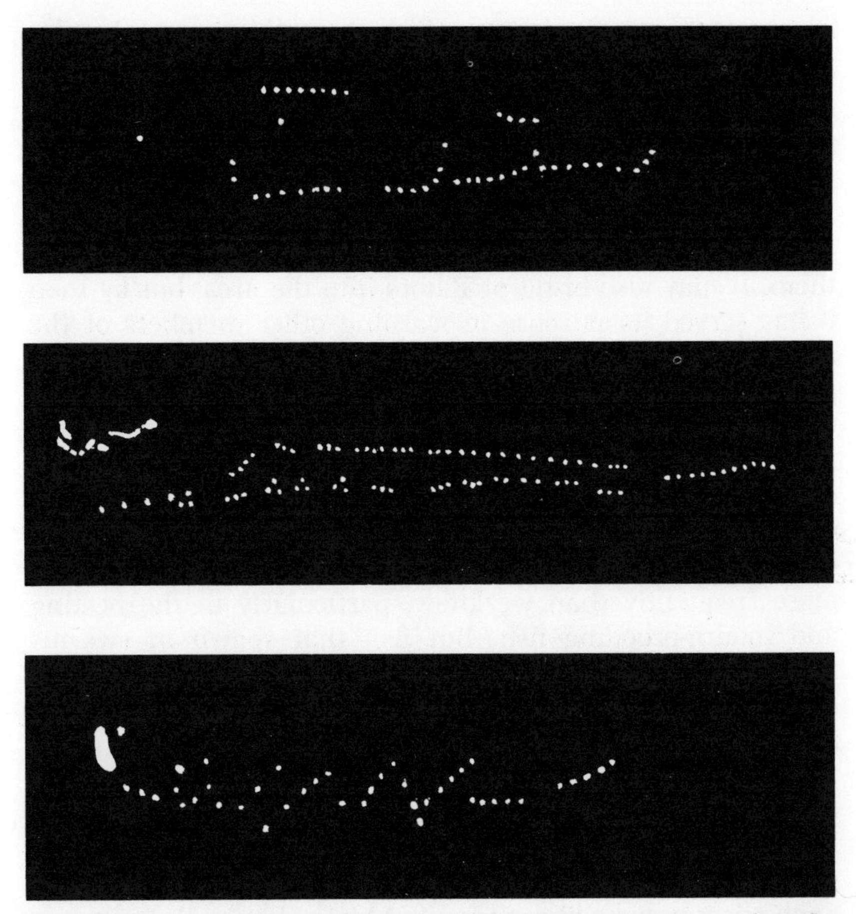

Figure 30. Countless deep-sea fish use light organs to identify their species and to signal each other. Every species has a different pattern, and sometimes male and female have individual patterns too.

Fish Using Odors. The use of odor for signaling is not as profitable in water as it is in air, except perhaps at very close range, because unless there is a good current to carry it odor not only travels more slowly in water, but it also becomes diluted and dispersed. Nevertheless, there are some uses for odor, and the fact that many fish have exceedingly well-developed smell areas in their brains and show great sensitivity to odor makes us less certain that odor signals are *not* used, but very little is known about them.

Many fish release a chemical into the water from their skin when they are alarmed or injured, just as we secrete more adrenaline when we are frightened or under stress. This chemical produces fright in other fish as its odor reaches them. It may also bring predators into the area, but by then it has served its purpose in warning other members of the species to beware and disappear. This is another of nature's provisions for the survival of the species; sacrificing one for the survival of many. It is not unlike the male sacrifice for female survival.

Of the very little we know about fish odors, we are certain of only one or two facts. Certain species of fish use a chemical for recognizing their own brood of young. This may be done more frequently than we know, particularly in the nesting and mouth-brooding fish. But fish that spawn in swarms never see their young, so it would not be expected in them.

There are species of fish living at the bottom of the ocean in which the female releases a chemical signal into the water when she is ready to spawn. When this signal reaches any male of the species, he follows it to her and mates with her. But such a chemical must often miss its target because of the few fish inhabiting some zones of these deep waters. Such odors may be like many used by land animals; only one species is sensitive to them, and other species and predators do not recognize them for what they are.

Sharks have particularly keen noses and can track down an odor with great ease, but certain odors have much more effect on them than others. These odors, like sharks' tracking ability, are involved only with food. So far there appears to be no evidence that sharks use odors for species identification or for attraction, much less for signaling their needs or their intentions.

SMELL WITHOUT A NOSE. Because it is more difficult to judge the direction of an odor in water than in air, many fish have developed odor detectors on the body surface. These detectors give fish much greater accuracy than a nose because unlike a nose, which must be pointed in all directions until the odor seems strongest, these body cells are already pointing in all directions. The fish's brain need only recognize the positions of the cells that are being affected the most.

Body surface cells seem to be very highly developed in eels, but it is not known if this has some special function for eels besides recognition of species, the sexes, and the detection of prey. The eel is a migratory fish, traveling great distances to its spawning grounds, and it may be possible that faint odors play a part in this migration, just as they have been proven to do in migrating salmon. If this is so, the special cells on the surface of the eel's body may be important in making such long migrations possible.

10. Using Electrical Currents

Several hundred species of fish have special organs for producing electricity that can be transmitted. They also have special centers in their brains for receiving and decoding electrical currents picked up by sensitive receptor cells in the skin, but no land animals have this ability. The use of electrical currents for any purpose that requires living cells to receive them must be confined to water, because water is a much better conductor of electricity than air is and only in water is it really effective.

The use of electrical currents of considerable power for protection against enemies and for stunning or killing prey is well known and was even recognized thousands of years ago in species of catfish, rays, and stargazers. But we know very little about communication by electricity except that it is used by many species for the detection of prey and to identify themselves in the search for a mate.

The manner in which this electricity is manufactured is not quite the same for both purposes. For communication only, a mere fraction of a volt is required, whereas fish that use electricity for defense and for killing prey produce up to 550 volts or more. The minute currents that make identification, detection, and attraction possible in muddy or dark waters are created by nerve and muscle action in many fish, but these fish are not always able to detect such currents. The currents are, in fact, not specially produced; they are natural to all muscle action, as will be explained shortly. The power of these currents is seldom more than a thirtieth of a volt, and they can be detected only by specially equipped fish.

Electrical Glands. One of the best known of the freshwater fish that uses batteries for transmitting electrical energy at a more powerful level than mere muscle action is the elephant fish [80] (shown in Figure 31), which can be used as an example for describing this process. The organs that manufacture the current are composed of a series of closely packed, disklike *electroplates* developed from muscle cells embedded in a jellylike substance. There are from 150 to 200 of these plates in each of four electrical organs situated near the tail, two on each side of the body—that is, from 600 to 800 plates altogether.

Batteries used for the "stunning" high voltages are situated quite differently, and we are not concerned with them here except to mention that the electric eel [40] has both these and the weaker ones just described for communication. The fish that use these batteries vary greatly in the frequencies of the pulses they discharge—the rate varying from one to a thousand per second. It is probably the pattern of this pulsing that identifies a species. We still have a great deal to learn about this coding and how it is used, but one thing has been shown by experiments. The electrical field a fish creates around itself with its electrical glands is

Figure 31. The elephant fish [80] creates quite a powerful electrical field for testing its surroundings and keeping in touch with other members of its species.

interfered with by solid bodies such as rocks or other fish, so the field can be used like radar for detecting and avoiding, for moving toward objects in dark and muddy water, and for species recognition by the pulse rate.

For an electrical field to be most efficient when it is used in this way, the animal's body must be rigid. Thus, we find these fish holding themselves quite straight while swimming, propelling themselves only with a long, specially developed anal or dorsal fin which is waved along the body to produce movement. But if immediate escape becomes necessary, normal and vigorous body movement will quickly take over.

Electrical Fields. Far from being surprised at the use of electrical transmission in fish, we should wonder why more fish have not chosen this way of detecting prey and communicating, because all nerves in a living body function as a result of electrical stimulation. Nerve fibers are actually specialized electrical conductors. When any sense organ is stimulated, a special pattern of electrical impulses is created that pass along nerve fibers to the brain, where they are translated into their sensations.

When a brain decides on some kind of action, it produces electrical current that passes along other nerve fibers to muscles, which then contract to move a limb or a part of

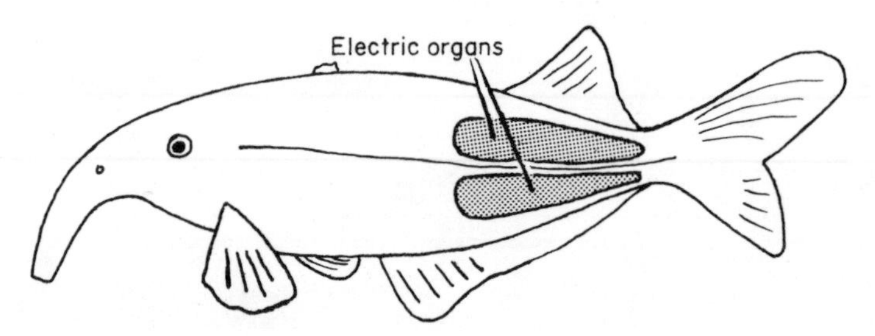

Figure 32. The electrical organs used by the elephant fish [80] and many others are glands situated close to the tail, two on each side of the body.

the body. The action of glands is also stimulated this way. All movement is therefore the result of electrical currents, and all currents create fields that can be detected by fish equipped with receiving organs for this purpose.

The creation of these fields is due to the fact that all electrical conductors have a certain amount of leakage into their surroundings and in the case of nerve fibers into their surrounding tissues. This can always be detected and measured by sensitive instruments. It is because of this leakage that we can use instruments such as the electrocardiograph and the electroencephalograph. They may seem impossible words, but "electrocardiograph" is only a combination of the words "electricity," "heart" (*kardia* in Greek), and "graph" (from the Greek word for "drawing"); and "electroencephalograph" is derived from "electricity," "brain" (*kephale* in Greek), and "graph."

The electrocardiograph picks up the electrical field created every time the heart muscles beat. By moving a pen on a special roll of paper, it transforms this into a graph that is different for every different way in which the heart muscle can act. The electroencephalograph detects how the brain is working and draws this on paper in the same way. Another instrument that acts in the same way is the lie detector.

To summarize these methods clearly, we can say that the electrical fields given off by muscle action can be detected by some fish with special receptors (just as we detect them with special instruments), and certain other fish can transmit another kind of minute electrical current themselves with special glands, the currents being received with special receptors by other members of the species or as radar echoes.

Some Effects of Electrical Fields. Many fish, especially sharks and rays, are so sensitive to the weak electrical fields just discussed that their breathing is even affected by them and the very weakest current slows the heartbeats of some of them. This tells us just how sensitive the receiving organs on their bodies must be—so sensitive that they will register currents as low as ten-millionths of a volt and so sensitive that the breathing rate of some rays will change if they receive a hundred-millionth of a volt. Both these levels are less than the current generated by the movement of a fish's gills.

Just being sensitive to electrical fields does not require the precision that is needed to locate their sources. Experiments have proved that some skates can detect other small fish buried in sand just by the electrical currents created by the movements of their gill muscles. Some deep-sea fish are just as sensitive even if they are not close to the bottom. They can detect other fish swimming at distances up to sixty feet.

For fish to be able to identify the source of an electrical current in the natural state, where countless such fields mingle with each other all around, there must be an efficient filtering system in the brain, as was suggested for sound and smell. Some of this filtering may be done in the receiving organs, for certain fish are sensitive to one level of current and others sensitive to other levels. But most filtering must take place in the brain.

The Receiving Organs. Very specialized areas in the lateral-line system of fish seem to be responsible for receiving electrical impulses. This is logical because sounds of certain wavelengths and sensitivity to current pressure and temperature are all registered there by separate receptors in the system. In sharks the organs sensitive to electrical currents are small skin pores that open into a fluid-filled canal behind them, where sensitive cells have tiny, hairlike cilia projecting into the fluid. It is also possible that the part of the lateral line in the region of the head is the most sensitive because the canal divides close to the eye to pass both above and below it.

The Brain Centers. All these fish have large, specially developed areas in the brain, where they interpret electrical signals and identify objects. The nerves to their lateral-line organs are also immensely developed compared with those of other fish. The nerves to the eyes and nose are, on the other hand, greatly reduced. The way this brain area enlarges more than any other sense area in fish using electrical communication is shown in Figure 33.

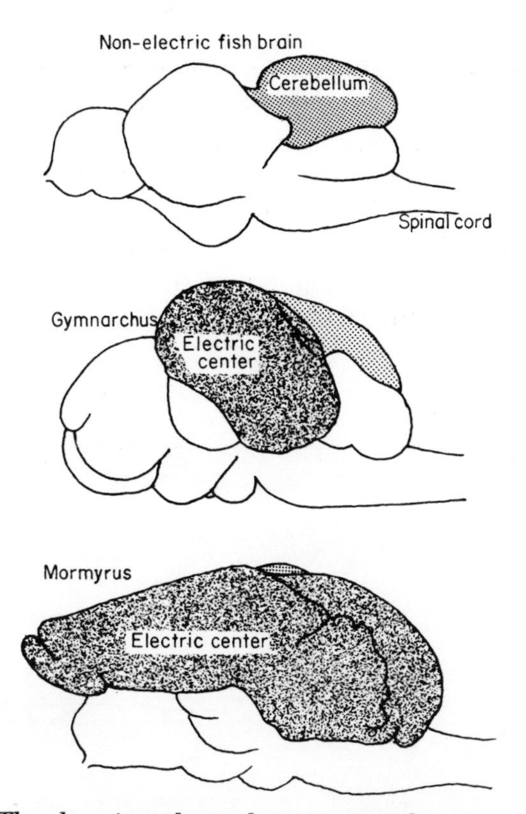

Figure 33. The drawing shows how a special center in the brain develops more than any other area in fish using electrical communication. It almost completely covers the visual and hearing centers.

Earlier we mentioned two important features of this electrical sensitivity: the extreme sensitivity of some fish and the need to filter out all unwanted fields. That this is done successfully seems rather incredible, but it seems even more incredible when we consider some recently discovered information. One fish [47] appears able to detect a minute change in the electrical field of the water around it—a change of only three-billionths of a volt per square millimeter. The electric eel [40] can also detect and judge the size, shape, and position of an object or prey at distances up to twenty feet.

Perhaps weak signals can be separated from strong ones at the lateral line, but it seems certain that all kinds of identification—including the fish's own radar echoes and their separation from other currents—must take place in the fish's remarkable brain center. Finally, in addition to all these tasks the fish must be able to identify other members of its own species and members of the opposite sex. All this without vision is truly remarkable.

11. *Sound and Other Signals in Amphibians and Reptiles*

Amphibian Amplifiers. Not all amphibians have voices, although most probably have. Those that do not have voices communicate with chemical, touch, or visible signals. This is not surprising, because amphibians are the lowest of the vertebrates to have larynxes and vocal cords. Those amphibians without tails (the anurans) definitely use voices for communicating, but many of them can make only a few sounds. Many of them, however, have apparatus for amplifying the sounds they make, as well as true but simple larynxes. Although others, like the caecilians, have larynxes, they can only make creaking and chirping sounds. Such sounds mean little to us, but they evidently mean something to other members of the species.

It seems that in many species of amphibians that use sound signals, only the males use their voices. The females also have larynxes, but these are less developed than in the males. There are certain species in which the females make quiet sounds. The female of the midwife toad [9] is quite exceptional in having a louder voice than the male.

Many male frogs and toads have large inflatable throat sacs; some species have two—one on each side of the lower jaw—and these can be internal or external. The throat sacs are used to amplify their calls, and they produce a volume of sound that will carry for a great distance. The air is passed from the lungs across the vocal cords in the larynx, then passes through slits in the floor of the mouth into the sac or sacs. Because the air must go out the same way it goes in, a repeating call can be made by keeping the mouth closed and forcing the air back and forth from the lungs to the vocal sacs. This not only amplifies the sound but also allows the frog to call or make sounds under water, where opening its mouth would not be possible.

Species that live near permanent water can breed at the same time each year, but those that have to breed quickly

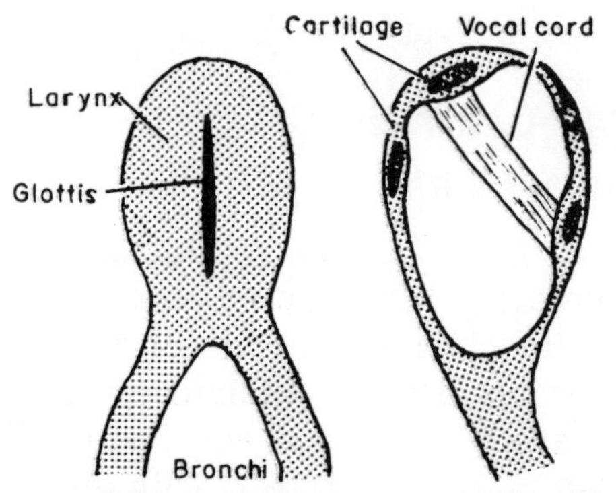

Figure 34. The amphibian larynx is a simple affair located between the entrance to the throat and the bronchi to the lungs.

when pools form after rain cannot be so regular. These amphibians have louder voices to be sure of attracting females quickly from a wide area so the full cycle of breeding and metamorphosis is possible before the pools dry up again. Species of small frogs have much shriller calls because they have much smaller larynxes.

Figure 35. Many male frogs have large throat sacs that resonate their sounds, the air being forced back and forth between these and the lungs.

Figure 36. Some frogs have two resonating throat sacs.

Amphibian Ears. A number of species of amphibians have no *tympanic membranes,* as their eardrums must be called. Those with tails [115] have no middle ears either, but they all have inner ears that can receive vibrations from the ground or through their bone structure. When amphibians do have tympanic membranes, the membranes are not always obvious. In many species they are covered by skin, but in others they can be quite prominent (see Figure 37).

When amphibians first came onto land, their hearing apparatus was very similar to that of fish. Then a gill chamber was adapted to end against the surface skin, where it formed a tympanic membrane right on the surface of the head (see Figures 37 and 38). There are several forms of this in amphibians. In one frog [99] this drum has retreated down an external chamber, like that of the ears of higher vertebrates. But some, such as the American tailed frog,[18] still have very rudimentary ear ossicles and no eustachian tube connecting the middle ear chamber with the throat.

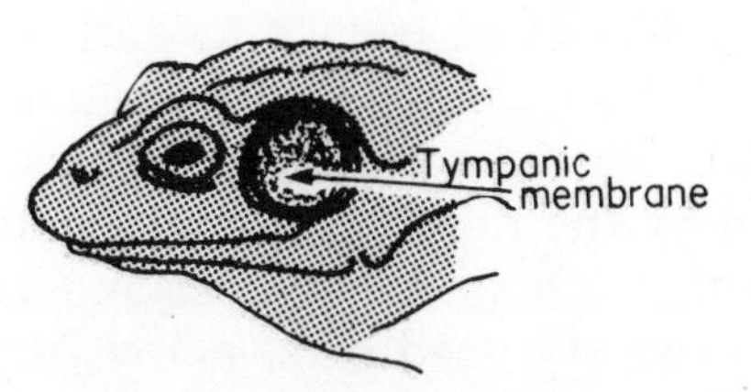

Figure 37. Many amphibians have a surface membrane close to the eye, which acts as an eardrum and transmits vibrations through a *columella* to the inner ear (see Figure 38).

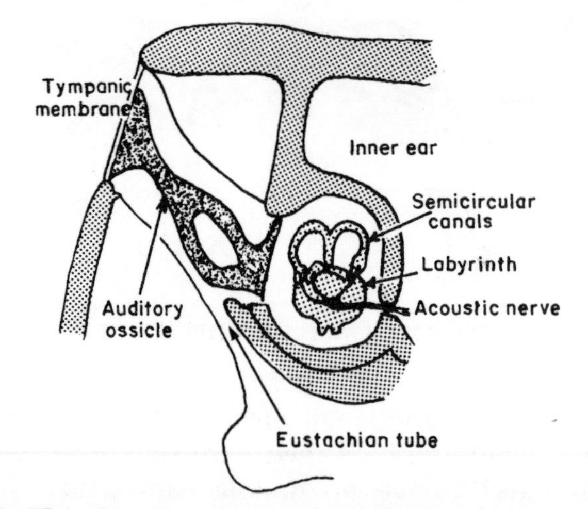

Figure 38. This drawing shows how the amphibian ear has only a single ossicle, the columella, as described in Figure 37.

Most frogs that use sound signals do more than just croak; they have many notes—some of which may be harsh but many of which are quite melodious. The notes range from the bass bellow of the bullfrog to the tiny chirps of tree frogs and from single notes to complicated rhythms and tones. Frog calls are considered to be mainly for attracting mates, but there are also calls for territorial defense, for warnings, and for gathering numbers of males together at a good breeding site where there is plenty of water for the larvae to develop. Male frogs remain in one place, and the females roam, like female crickets and most birds. This is why male frogs use special mating calls to attract females to their ponds.

Many different species of frogs may gather to mate in one area, even in the same pond, so sound signals must be very precise to attract the right females. Species calls are identified quite often by the number of times notes are repeated every minute, the pitch of the notes, and their length. In spite of having not very good ears, many frogs can respond to calls within a range of 30 to 15,000 cps., and some can recognize their species call up to a third of a mile away.

Besides mating and other calls previously mentioned, frogs have distress calls and so-called release calls. The latter, usually a loud squawk, is uttered by a nonbreeding female or another male that has been grasped by mistake by a breeding male. A male may also utter a release call after mating.

The females of many species have no voices, but species that do produce sounds usually have only croaks or screams, not calls. There are some exceptions to this. Both sexes of reed frogs [5, 54] call to each other, but the male is able to call more loudly because it has a large throat sac below its chin that amplifies sound. With some two thousand species already identified it would be strange if there were not some frogs and toads that are silent. There are a few species that either have no middle ear or that live in noisy areas where rushing water drowns out all sounds that could be of use to them. Calling is naturally useless to those that cannot hear, but some make feeble squeaks when they are below water.

Not all frogs call singly; some call in duets, trios, or quartets, each with a different note. But there is such a maze of sounds at times that it is not easy for us to recognize these calls. Groups of frogs of the same species that become divided by geographic obstacles develop slightly different voice patterns and calls or different dialects, just as other animals do.

Amphibians that live entirely in water can be expected to be less able to call than amphibians that live wholly or partly on land, like frogs and toads; and it is difficult to find records of sound signaling in most of them. However, there are salamanders that possess vocal cords and make faint sounds. Even newts and one species of salamander without lungs [12] can utter slight squeaks.

Large salamanders such as the siren and amphiuma are said to make a whistling sound, and one giant salamander called *Megalobatrachus* makes a shrill call. Some others are said to make clicking, clucking, or kissing sounds by taking air into suddenly opened mouths; but all these sounds may be the result of handling or alarm and only due to the expulsion of air or its inhalation. We have no real knowledge of sound communication in salamanders.

Reptiles. Snakes are voiceless, but the rest of the reptiles are not, although little is known of the signals most reptiles make. Alligators and crocodiles probably make more noise than the rest of the reptiles together. The males bellow loudly in the spring when the mating urge is on them, and they make a short, loud alarm call when danger is sensed, at which all within hearing distance rush for the water and submerge. Both sexes also hiss with an expulsion of air when uncertain about the approach of a possible enemy.

When young alligators are hatched from their eggs, they squeak loudly, and when the female hears this, she digs away the nest mound to let them out. The young stay around her for several weeks, croaking and grunting in a conversational way, but this is probably nothing more than recognition or feeding signals.

The next-loudest voice to the crocodiles or alligators is probably that of the Galapagos tortoise. The male's mating call is a deep, hoarse roar. Other land tortoises have voice sounds that range from a loud, repeated, rasping noise to a mere threatening hiss. Apart from defiance or threat it is difficult to know what their sounds mean.

The noisiest group of lizards are the geckos, which do not confine their calls to mating alone even though they make

Figure 39. The giant Galapagos tortoise [107] roars at a rival, which keeps its head well tucked in away from attack.

few distinguishable sounds. It is understandable that these animals should call because so many of them are nocturnal. Lizard sounds that have been identified are calls for defending territory, barks for threatening an enemy, and squeaks when disturbed. One lizard [49] has a victory cry after defeating an opponent or catching prey. Another [94] will utter a prolonged scream like that of a human child if it is disturbed.

Most of the legless lizards have feeble voices. Burton's legless lizard [61] makes a squeaky sound that is rather drawn out. The sounds of some other species are like those of the quieter geckos. Although many lizards do make sounds, a number of them do little more than hiss. On the whole we know very little of how lizards communicate with each other.

RATTLES AND HISSES. Snakes do not hear as other vertebrates do; their ears are so rudimentary that they can hardly be sensitive to sound, although they are sensitive to vibration through the ground. When snakes degenerated from lizard forms, they gave up more than their legs. They also sacrificed their ears, so one hardly expects to find snakes making sounds for signaling, yet occasionally they seem to.

The rattlesnake naturally comes to mind, but to what extent this animal makes its rattling sound intentionally is not certain. The sound is produced by the movement of several buttons of horny skin that are loosely held together on the end of the tail and which are added at the rate of two or three a year. Because they break off occasionally, no more than a dozen rattles are found on any one snake.

The older and bigger the rattlesnake is, the louder its rattle is likely to be. It is doubtful that the rattle is really a true signal; it is more likely the result of agitation or tension. Nevertheless, the rattle acts as an effective warning.

The hiss of a snake is merely the expulsion of air or the drawing of air into the lungs. It may only be the result of alarm, but it acts as a threat signal. Some snakes have developed a membrane in the throat that helps break the hiss

Figure 40. The rattlesnake's rattle is a series of buttons of hard skin which grow at the rate of two or three a year, but which fall off so that there are seldom more than nine or ten. This is an excellent specimen (inset) with twelve buttons and showing where one has dropped off.

into a string of sounds. In the large African puff adder [23] it produces a snort similar to that of a horse.

The rattlesnake's rattle is the loudest kind of snake noise, but certain other snakes certainly have similar warning reactions, even though they have no rattles. In these other species the vibration of the tail merely produces a buzzing or rustling sound, which can be heard up to fifteen feet away if the snake is on dry leaves. The American copperhead,[6] the bushmaster,[57] and many other tropical vipers have this habit, which is often very fortunate for man but is hardly communication between members of the species.

We may be correct, however, in accepting the rattle of the rattlesnake as a signal of "Keep away, or I'll bite," because if it were biting prey, it would not rattle. Perhaps it rattles when it is approached by something so large that it feels in danger—something like a man, a horse, or a mountain lion. This is like a little man shaking his fist at a big man he knows he cannot beat in a fight.

Figure 41. Reptiles' ears, like those of birds, have no pinnas, so they do not collect as much sound as mammals. This picture of an estuarine crocodile shows the entrance to the ear canal, which can be closed by a flap of skin. Below, other aquatic reptiles have similar ear structures.

OTHER REPTILE SOUNDS. Few sounds are made by other reptiles but all these sounds originate from the lungs and throat, except in one species of gecko that rubs plates of scales together to produce a harsh, creaky sound. In spite of this the ears of lizards are well developed and efficient—an effective aid for survival, even if communication is not at a high level. These ears are often covered by scales (see Figure 41), but many have openings to the outside like the higher animals. In fact, all reptiles except snakes have a tympanic membrane, ossicles, and a middle-ear chamber as well as an inner ear. In addition crocodiles have a prominent ear, but it is covered by a flap of skin. Most reptiles, therefore, must have fairly good hearing.

Color and Other Signals. Like fish many amphibians use color for camouflage, but only a few appear to use it as a signal. Some use posture, and others use odor or touch in one way or another. The colors are derived in the same way as those of fish—from pigment cells and *guanophores* or iridocytes. In some the color change can be almost instan-

Figure 42. The iguana has a large surface scale over the site of each ear.

Figure 43. The leaf-tail gecko [94] signals both visually and with sound, although the amount of communication is limited in both methods.

taneous. This color change can be prompted by what the animal sees; this has been proved by removing the eyes and finding then that no color change takes place, except that some cells change a little as light changes to darkness.

Most of the change is camouflage and takes place at any time, but there is also another response to gland or nervous influence that brings about more rapid changes in excitement, fear, or anticipation. Perhaps none of these changes are pure signals—except that which is seen in the breeding season, when the males of some species take on special colors. But the sudden revelation of hidden brilliant colors on the belly or inside the legs may be a signal that the owner is distasteful or even poisonous to any intending predator, or it may distract the latter long enough for the owner to leap out of reach. Except for some tree frogs, few of the color changes of amphibians are as rapid as is possible in many fish, nor are they as obvious.

There are definite color differences in the sexes of European newts and some other species of amphibians, but the actual signal for breeding is the release of secretion from a skin gland that has a stimulating odor. This seems to be true with salamanders even when they also use display. A few frogs use odor to keep groups together and color or body movement to attract females, but frogs and toads as a whole rely so much on sound signals that there is little more we

Figure 44. The Australian frilled lizard [29] is shown here with its frill folded back. When spread in threat or defiance, this makes the animal appear several times its normal size.

can say about color. Unless we just consider breeding behavior, there is just as little to be said about posture.

Lizards are a very active part of the reptile world that have well-developed senses, and many species have unmistakable signals, apart from their sounds. Many geckos, for instance, have a definite alarm and threat signal, in which they raise themselves right up on the tips of their toes, inflate their bodies with air, and squeak or bark as they launch themselves an inch or so toward the creature confronting them. In the same circumstances some others release a skin secretion with an unpleasant odor.

Many of the Australian dragons have an interesting habit of waving a front leg in the air while bobbing their head. It is assumed that this may be a signal because both sexes do it, but its purpose is yet unknown, although it has been observed as a preliminary to mating. Head-bobbing alone is known to be both a mating and a territorial signal, and the speed of bobbing—like the speed of push-ups by some other lizards—varies with the species.

Many lizards use changing colors for camouflage, but in some species the males may also change color in the mating season and display warning colors when threatened. Chameleons and iguanas are active color-changers. They react very rapidly to excitement, but they also change color with temperature, so it is difficult to be sure of the significance of any reaction except camouflage. Threat signals are very obvious, however, in species such as the Australian frilled lizard,[29] which opens its mouth, spreads a large neck frill, and hisses, and the bearded lizard,[11] which does the same things with a spiky beard and a pouch while inflating its body with air.

There are lizards that open their mouths when they are threatened and show brilliant colors inside. This is a threat signal which, although it could seldom be followed up with action, is often effective in saving the lizards from attack. As far as we know, lizards that are active by day use visual signals to communicate, and those that are nocturnal employ odor or sound.

If there is little to say about lizard communication, there is less to say about that of snakes. This may be largely be-

Figure 45. The bearded lizard [11] inflates its scaly beard pouch and body and opens its mouth wide to signal threat or attack.

cause of lack of interest in studying snakes; most attention has been given to studying the more dramatic killing power of snakes.

Snakes are presumably insensitive to sound, and the vision of most snakes may be less acute than that of lizards. Their sense of smell, however, is exceedingly acute due to the presence in the roof of the mouth of Jacobson's organ, into which the tongue flicks particles of scent picked up from the air. It is possible that because of this organ snakes use odor signals for attracting mates, for identification, and for leaving trails. It is known that the female garter snake [109] releases a sex pheromone to attract males, so other snakes may do the same.

Snakes use color quite extensively for camouflage, but a few also use it as a warning signal of their deadliness. They make themselves thoroughly conspicuous. The coral snakes [78] do this and are carefully left alone by predators who recognize their bold patterns of red, yellow, and black bands. In fact, so effective is this warning signal that wherever any of the species occur, other nonpoisonous snakes can be found with similar colors and markings, taking advantage of the protection from attack this gives them.

Apart from this kind of warning signal, perhaps only that of the cobra is well known. Its habit of spreading its neck into a hood, on the back of which a pair of large eyelike markings appear, is not only a threat signal but is calculated to intimidate either enemies or prey.

Of all the reptiles perhaps least is known about signals among tortoises and turtles. Apart from the sounds mentioned earlier, odor seems to be their only medium for messages. Musk and mud turtles produce a glandular musk odor when alarmed or giving a warning, and in time it may be discovered that other tortoises and turtles do this too. Certainly some use odor to attract the opposite sex. On the whole we have a great deal to learn about these almost prehistoric creatures.

PART V
COMMUNICATION IN INSECTS AND CRUSTACEANS

12. *Sounds and Visible Signals*

Sounds. For their size some insects might be considered the noisiest creatures on earth. Fortunately for us, only a few species are capable of making great noise within our hearing range, either singly or in chorus; and even more fortunately, insects are small. If a cicada were the size of a dog, for instance, it would put a supersonic plane to shame with its noise.

MECHANICAL INVERTEBRATE SOUNDS. Everyone is aware of the sounds made by locusts, crickets, katydids, grasshoppers, cicadas, some beetles, bees, mosquitoes, and other insects. In some of them, such as bees and mosquitoes, it may only be the frequency of their wing beats that produce their sounds. Although this can hardly be called a signal, it allows members of a species to recognize each other and sometimes to recognize the opposite sex. In most other insects sounds are made only by males. This is true in crickets and grasshoppers, which have various sound-producing devices, most of which function by what is called *stridulation.*

One scientist has recognized twenty-one different methods of producing stridulation by insects—for instance, four in butterflies and moths, five or more in spiders, and four in scorpions. Most of the methods work on very similar principles.

In some grasshoppers stridulation is produced by scraping rows of tiny teeth called a *file* or *pars stridens* (located on the inner sides of the thighs of their back legs) across the sharp edge of a wing vein called a *plectrum* or *scraper.* This makes the rasping sound we know so well. This is a

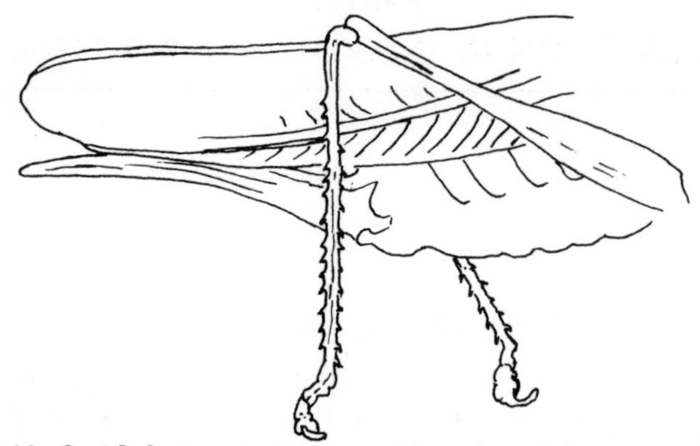

Figure 46. Stridulation is the sound produced by many kinds of grasshopper, cricket, etc. It is made by scraping rows of tiny teeth on the inner thighs of the hind legs (called a *file*) across the edge of a wing vein (called a *scraper*).

love call, and except for an alarm call if it is caught or picked up, this may be the only call some species of insect make. Variations of this method of producing stridulation are found in most grasshoppers. Some, however, grind their teeth together to make their song.

On the whole grasshopper sounds have a very wide range of frequencies, some going far beyond the limits of the human ear. For instance, the large European green grasshopper [108] can produce frequencies as high as 100,000 cps. This is 80,000 cps. above the upper limit of our hearing. Other scraping devices used by insects are found on the head, the thorax, the abdomen, the legs, the clytra, or the forewings.

Katydids have a wide range of song notes. In the male of one species the left wing has a small filelike edge to a vein. The right wing has a ridge over which the file rasps when with the left wing overlapping the right, the wings are moved sideways, and the thin membranes near the roots of the wings are vibrated. Each species has its own tune. Some of the katydids have one note for bright days and another for night or dull days, but the reason for this is not known.

It might seem at first that most insect sounds are associated with mating, but many are also used for other kinds of signals. Male crickets, for instance, have sounds for calling each other, for courtship, for copulation, for courtship interruption, for aggression, and for recognition. Some insects go further: they even have calls that mimic their prey —to make capture easier.

Large gatherings of cicadas in the mating season can make an unbearable din, which apparently serves the double purpose of attracting the opposite sex and driving away birds and other likely predators that cannot stand the noise. The frequencies used by the various species of cicadas range from 600 to 16,000 cps. The seventeen-year cicada [68] makes its sounds in the range of 900 to 1,600 cps., although it can actually hear in a range from 600 to 3,600 cps. Another species of cicada [66] uses a higher range—from 4,500 to 6,000 cps.—and it too is able to hear over a much wider range—from 600 to 10,000 cps.

Some species of cicada may tune their songs to coincide with a certain temperature or level of humidity, because they often take their frequencies up and down two or three times before settling on a definite note. They may also try more than one note. When cicadas are picked up, some species give off short, sharp calls, which may be the same as a frog's release call, but too little is known about it to be certain.

The males of some seventeen-year cicadas [67,68] produce their sounds by buckling one after another a series of stiff ribs buried in a flexible body drum. Each of these movements makes a resonating chamber vibrate ten or twelve times, so there are as many as a thousand vibrations a second. Sometimes called a song, this sound can be used to identify different species. It will also attract the opposite sex and warn off predators.

Other insect sound signals are made by tapping the head or abdomen on the ground, vibrating the antennae, snapping the mandibles, or vibrating the wings while at rest.

HONEYBEES. These insects have quite an extensive sound vocabulary because they make different sounds in so many situations. For instance, the queen bee "pipes" at about 500 cps., sometimes so loudly that it can be heard ten or twelve feet from the hive by a person with good hearing. If the queen leaves the hive to establish a new colony, an unhatched queen starts piping, or *quacking,* in her cell at a lower frequency until she is released by other members of the swarm.

Considerable attention is given to the dance language of bees later in the chapter. This dancing is accompanied by sounds at a frequency of 250 cps. when the direction of food is being indicated in the dance. It has been discovered experimentally that the average length of the sound and the number of pulses in this dance are directly related to the distance the bee has traveled from the food supply it has found. This suggests that the bee communicates an idea of distance through its sound language. The number of pulses in the sound may also tell how concentrated the sugar is in the food discovered. This sound language must be very important because other bees will not leave the hive without it; the dance alone is not enough. Altogether bees appear to make ten different kinds of sounds within the hive.

MOTHS AND ULTRASOUND. A device is used by certain moths to produce ultrasonic sounds that can be heard by bats hunting them, just as the bat's ultrasonic sounds can be heard by the moths. But the use of these is not yet fully understood. There are, however, two clear findings from experiments: the moths also give off these sounds when they are touched, and bats swerve away from these sounds. Perhaps, therefore, this signal is taken as a warning by a bat, which will act as though the sound is being made by another bat.

If this is so, these moths have learned to protect themselves from bats by imitating them. This is very much like the way many harmless snakes protect themselves by adopting the same colors as the venomous coral snakes—protection by mimicking. Both sexes of the death's head moth [2] can make a loud sound by passing air through the pharynx (throat), and this sound can be made to throb by alternately taking in and expelling air.

This moth usually makes two very short sequences repeated rapidly, one low-pitched and one high-pitched. The low one is made by dilating the throat cavity, which produces a breath of air through the proboscis. This causes a crescent-shaped clapper called an *epipharynx* to vibrate. The high sequence is produced by the expulsion of air, which creates a whistle.

Countless other sounds in the insect world are obviously signals, but we don't understand them. We don't know to what extent they may be nothing more than a reaction to chemical influence. For instance, soldier termites knock their heads against the wooden walls of their tunnels to signal, and the vibration is felt by other members of the colony through their feet. This is a quick way to get a message over what is a long distance for a termite. Since termites are incapable of thought and may only be registering a change of body chemistry due to excitement, there is no plan or purpose in the action. Nevertheless, it may be a true signal.

There are many other insect sounds that we do not understand. A creature called a pill millipede [17] produces vibrations and squeaks that must serve some purpose, but so far the explanation hasn't been found. Even though sounds like the vibration of wings may serve to keep a swarm of insects together or to identify species, sexes, or individuals, who is to say that this is anything more than genetic or even a response to temperature or some other not very obvious factor.

INVERTEBRATE EARS. No insects have true ears as ears have been described for vertebrates, but they do have hearing organs, sometimes called *tympanal organs,* and for convenience we sometimes call these "ears." Short-horned grasshoppers, cicadas, and some moths have ears on each side of the body; katydids, crickets, and long-horned grasshoppers have them in their front legs (see Figure 47). Locusts have large organs on the sides of their bodies, which

Figure 47. The kind of ear found in the tibia of an insect's leg, and beside it a magnified drawing of the sense cells. Below is a cross-section of the leg.

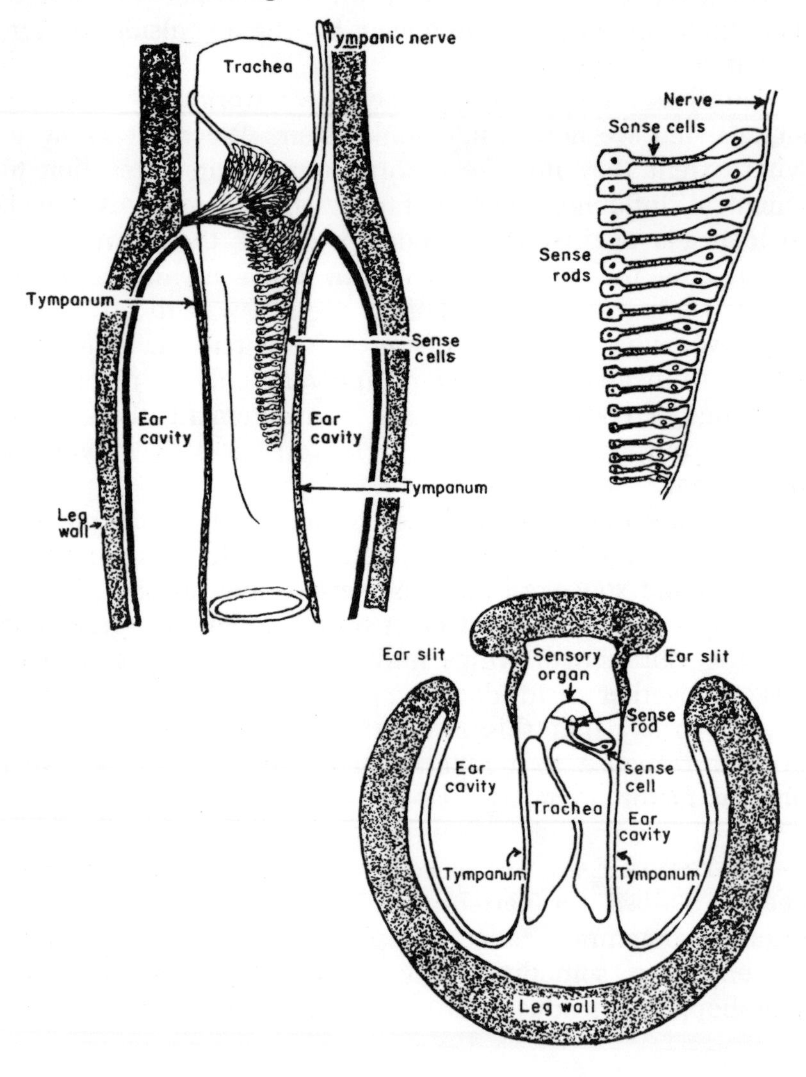

some experts believe are hearing organs, and they probably are.

Insect tympanal organs occur in pairs, each capable of receiving and locating sounds from any direction. The pairs all operate in the same way. A tympanic membrane or drum vibrates in response to sound waves and transmits the vibrations to a sensory organ behind it called a *chordotonal sensilla,* which transforms the vibrations into electrical messages to the brain. As simple as these ears are, they enable an insect to hear a much wider range of sounds in some cases than man. But it is possible that these organs respond more to loudness than to actual frequency or pitch of sound.

Worker bees, ants, and wasps have chordotonal organs on their legs, and these are sensitive to vibrations but less so than the ears of some mammals. Perhaps much of the usefulness of these organs in such insects comes from vibrations felt through their legs from whatever they are standing on. Bees may receive sound through their antennae too, because each antenna has a large number of plates like the tympanic organs of other insects, and these are thought to be sensitive to vibration. Similar receptors are found on the antennae of a number of other insects.

Night-flying moths [16,45,83] have tympanic membranes tuned to the ultrasonic sounds made by bats. This enables them to hear and evade bats that hunt them. These ears are located on the thorax, and although they are minute, they can be seen as small cavities containing transparent drums. As simple as this kind of ear is, it can detect pulsed sounds with frequencies from 10,000 to 100,000 cps.

It has already been mentioned that these moths can make sounds similar to those of bats. They have a row of fine ridges of skin that bend and spring back when a leg muscle contracts and relaxes rapidly, making high-pitched clicks up to one thousand a second in the range of the bat's ultrasonic sounds. The only possible explanation for these clicks seems to be to confuse bats by identifying with them.

Male mosquitoes can hear through their antennae while flying, and this is why they are able to keep together in swarms of males only. The male mosquito also identifies a female by her hum. It has been found that these insects are attracted most by sounds between 300 and 700 cps. By contrast, it is suspected that honeybees cannot hear at all while flying and so are not disturbed by the great hum that surrounds them in a swarm. But this contradicts what has been said about bees using sound to keep a swarm together, so it becomes obvious that even experts disagree on some of these things.

The field cricket has an interesting collection of "songs," as its stridulations are often called. They include a "calling" song, "courtship" song, and a "rivalry" song. Some species also have a "response" song between male and female. The sound may be of the same pitch for all of the songs, the difference in meaning being in the number of chirps each second. Only virgin females are attracted by a male's song. Once they have mated, the song has no effect on them.

Visible Signals. We are not sure of the full extent to which insects use visual signals, apart from the use of luminous organs, but they must recognize the shapes and color patterns of their own species, otherwise they would probably prey on each other. We know that many insects have excellent vision in their compound eyes of a kind that enables them to see and react to movement. This kind of vision must be highly developed in the fast-flying insects to enable them to avoid collisions and capture by insect-eating birds. We know that spiders make visible signals with their palps and legs, but it seems that most insects use sound, odor, or touch for communication. Some use light, but they appear to do this only for congregating together and attracting mates.

Like many fish, a few insects make visible signals at night with luminous organs. In daylight others use postures, sometimes color, patterns of movement, dancing (in bees), and flight patterns (in some butterflies). Although the color and flight patterns are only species and sex identifications, the dancing of bees is a really advanced form of communication, which will be discussed at greater length later in the chapter.

FIREFLIES AND GLOWWORMS. Relatively few insects use light organs, but those that do sometimes produce a most dramatic scene. They are all found in two families of beetles, fireflies [26] and glowworms.[58] The brightness of light they produce varies from a faint glow to quite brilliant flashing. It is doubtful if these lights are used for anything but mating, but turning them off may also be an alarm signal.

Of the various forms of luminous organs, most are groups of cells around the trachea and behind transparent skin, but some fireflies carry them on their tail ends. The manufacture of the light used by these insects is precisely the same as that used by fish, and it is just as efficient. But the organs that produce flashes are more complex and precise in action than those that merely glow. The males of most fireflies have large eyes with many facets which receive light from wide angles. Often the flies have light organs that point in several different directions, giving a species a very good opportunity for bringing the sexes together when they are scattered.

In some species only the males flash or glow. This is the same as in many other kinds of animals in which only the males call or show special colors. But there are a number of species of glowworms in which both sexes glow, the female usually replying to the male. The European glowworm [59] and a number of others operate in the reverse way, the male with only a feeble glow replying to the female's bright attracting signal, which she shows during certain hours of the evening.

Fireflies also have different ways of showing their lights. The males of one genus [93] flash, glow, or pulse their glow, and in some species the females also glow at certain times of the night. In other species they glow continuously. In still other species the male gives a double flash, and the female replies with a single flash. Male fireflies have large eyes which enable them to see and fly to the females as quickly as possible, so the lights need only be shown for the shortest time. This way fireflies are least likely to attract enemies and predators.

One disadvantage of flashing is that the sender must be more or less in line with the direction of view of the receiving insect at the moment of flashing. If any foliage intervenes, the flash may never be seen. It is also important that the receiver sees the flash and replies to it very quickly —to be sure that the two sexes reach each other. Sometimes

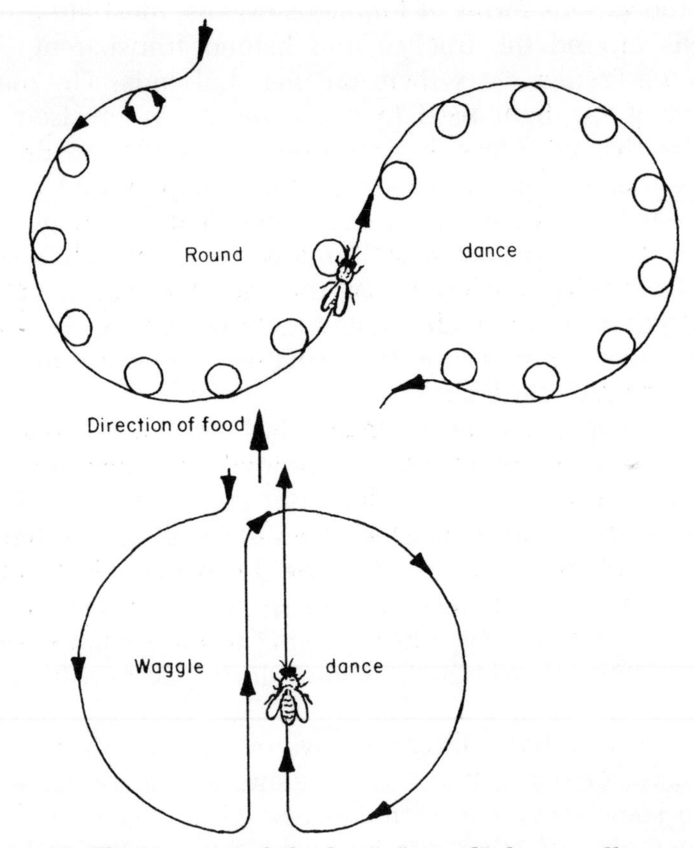

Figure 48. The pattern of the bee's "round" dance telling its hive mates of the discovery of extra good food nearby. Below is the "waggle" dance, which shows the direction of food at a greater distance.

a female response is only a third of a second after a male's flash.

There is very little uniformity in the way these insects behave, for we find another species [64] in which a female makes a flashing signal to attract males; as a male approaches making his own flashes, her flashing becomes more vigorous. But on the whole it seems that the patterns of flashing in these insects are designed to use as little energy as possible. But there are even exceptions to this. Some species in Southeast Asia collect together in millions, the males flashing together night after night for months. The fireflies are so densely packed in certain trees that these are called "firefly mangroves" and are used as navigation aids by river boatmen. The fireflies flash continuously when they are sitting in trees, but they seldom flash when flying —this is also different from many other species. Although many species of fireflies may be present together in an area, most do not mate in masses.

THE BEES' DANCE. A great deal has been written about the way in which bees communicate through dancing, and the more the dancing is studied, the more interesting it becomes. Since we know that bees also use sound, the complicated nature of the bees' language is quite astonishing. Bees are like people in a way; different groups have different habits and dialects, even though they belong to the same genus. Already several groups of bees have been found that have different dance patterns for the same purpose and that carry similar messages in spite of these differences.

There are two main kinds of dance used when an unusually good quality food (nectar) is discovered. One describes the location of the food when it is fairly close to the hive and the other when it is at a greater distance. Both the direction and the distance of the food are described in the second dance pattern, and the quality is conveyed by odor on the bee's body and by the sounds it makes.

Figure 48 (top) shows the "round" dance of a honeybee that has found a good supply of nectar within about ninety

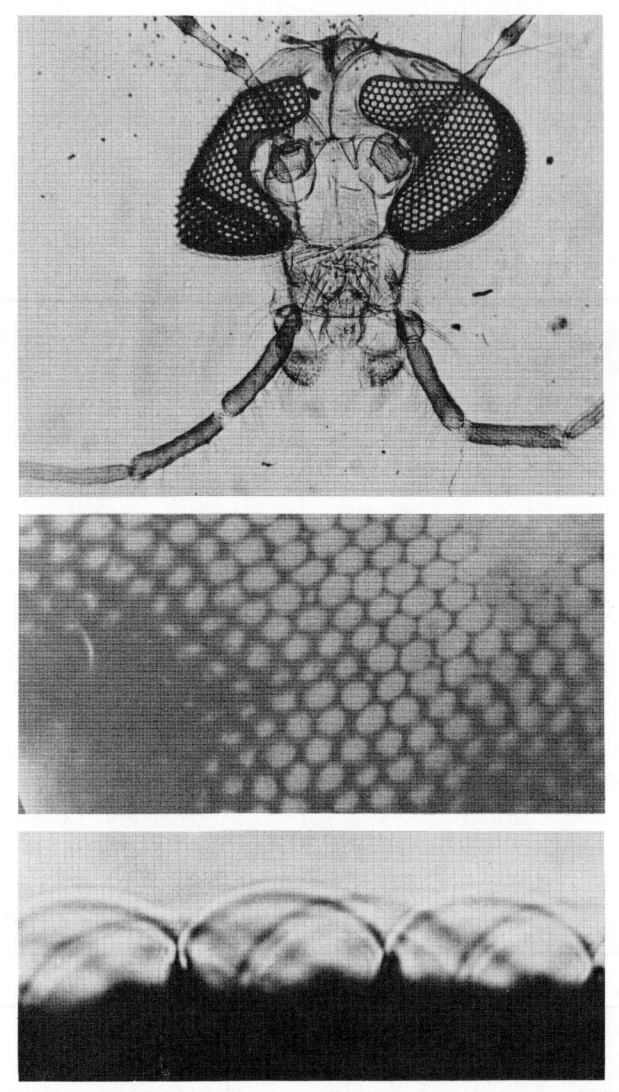

Figure 49. The compound eyes of the common housefly cover more than two-thirds of its head. The lenses of each visual cell can be seen on the edges. Center is a magnified view of the visual cell surfaces, and at the bottom a microscopical picture showing how highly curved are the lenses that focus the light into them.

meters, as described by Dr. Von Frisch. Figure 48 (bottom) illustrates a "waggle" dance, which tells of the position of food at a greater distance. The dancing bee follows the outline drawn, and as it goes along the straight central track that indicates the direction of the find, it waggles its body from side to side. It will repeat this dance until the message has been picked up by enough bees to make sure the nectar is well collected.

If a single pattern of the dance takes 1½ seconds to complete, the food is about 100 meters away; if it takes 2 seconds, the distance is about 200 meters, and if it takes almost 5 seconds, then the distance could be a mile. If the bee has to dance inside the hive because there is no convenient flat area outside, it may have to dance on a vertical surface, and the direction then has to be shown in another way. If the waggle is straight up, the direction to be followed is toward the sun; straight down is away from the sun. The number of degrees of angle away from the vertical indicates the number of degrees away from the sun, and allowance always seems to be made for wind whether it is ahead, behind, or from the side.

But what if there is no sun? A bee's compound eyes, like

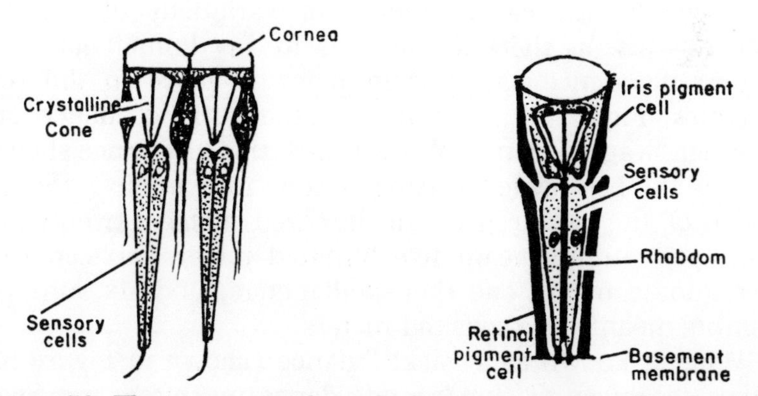

Figure 50. The structure of single facets or visual cells of insect compound eyes of the kind shown in Figures 49 and 51. The left one is from the group that includes silverfish, and the right one is a typical cell of an insect that is active by day.

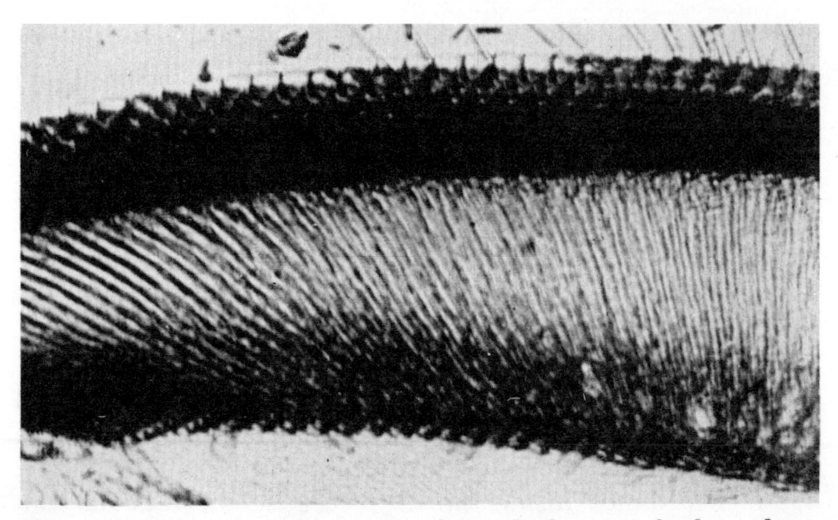

Figure 51. A microscopic section through the eye of a bee, showing the surface lenses, iris pigment, and visual cells.

those of many other insects, are capable of analyzing light and using what is called its polarity to identify direction. This can still be done even when the sun is over the horizon. If the sun is covered by cloud, the insect may still judge its direction by the intensity of the light, which will be greatest in the direction of the sun and least in the opposite direction.

As mentioned, earlier there are variations of the bee dances—just as there are dialects in any human language—and the dance varies from place to place in different colonies. Figure 52 illustrates another "round" dance and another "waggle" dance. When timed, this last dance showed that if the bee made thirty-eight runs in a minute, the distance of the nectar was one hundred meters; twenty-four runs per minute meant five hundred meters; sixteen runs per minute meant one thousand meters; and six runs per minute meant five thousand meters.

What is known as a "sickle" dance (shown in Figure 53) takes the place of the "round" dance in Dutch and Swiss bees, and this dance also shows direction. The more countries in which these observations are made, the more variations of the dance are found—but always with the same principles for describing distance and direction. Bees have a special dance when water is urgently needed to cool a hive, and they also have a particular dance when preparing to swarm to a new nesting site after scouts have inspected the surrounding countryside and brought back information.

Undoubtedly, the bees' communication system is highly complicated, combining visual signals in their dances with odor and sound signals. They recognize particular flowers by color, shape, and perfume—especially the perfume of the nectar. There are experts who feel this is not yet the whole story, and this is probably true.

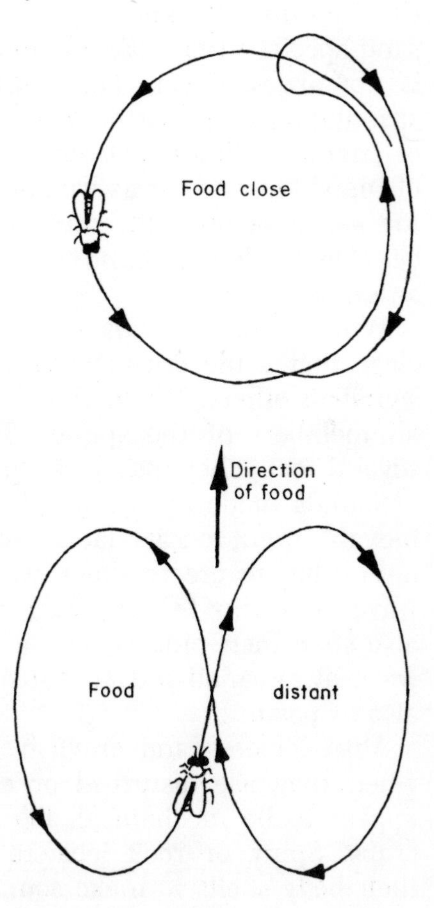

Figure 52. Another version of the bee's "round" and "waggle" dances.

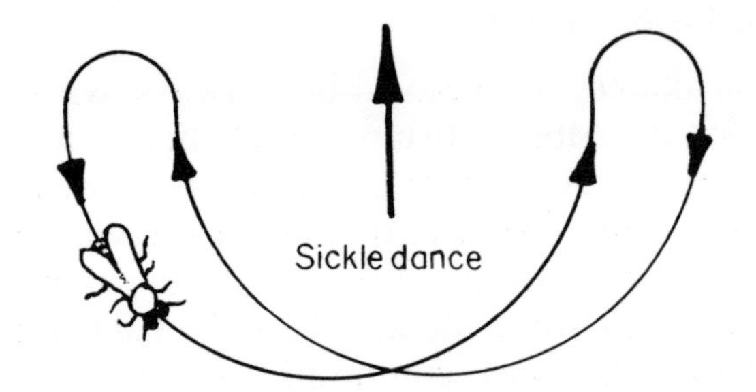

Figure 53. The "sickle" dance takes the place of the "round" dance in some colonies of bees, and it is unusual because it also indicates the direction of the nectar.

Crustaceans. We know that many of the twenty-six thousand species of crustaceans make sounds, but their purpose is not always clear. Those sounds that are not made by stridulation may just be body sounds due to the movement or friction of hard skeleton segments against each other or of mandibles. As far as we know, crustaceans have no hearing sense as we understand hearing. They may, however, be able to detect sound in other ways—perhaps through antennae.

Whether crustaceans have hearing organs or not, it is clear that some crustaceans can use sound as a warning signal to others. When this happens in burrowing species, all members of the species disappear within a radius of several feet. They may pick up vibration through their legs.

Sounds made by the pistol or snapping shrimp [8] with a movable joint on its large, single pincer are not for signaling but to create shock waves in the water. The shock waves stun tiny fish on which the shrimp feeds. Some crabs also snap their pincers, and it is possible that these sounds warn away small predatory fish. But many crab signals are also by posture.

Most lobsters and crayfish [89] produce fairly loud noises when they are disturbed or caught, but these sounds all appear to be mechanical, like the sounds of shrimps and crabs. Spiny or rock lobsters rub their antennae against their body shells to make sounds like a gate that needs oil, and many other crustaceans vibrate their claws. That these noises will be audible for some distance in the immediate surroundings is obvious because it has been said that a barnacle snapping shut its shell can be heard for ten miles. This is as difficult to accept, however, as it is to believe that crustacean sounds have any real communication purpose. But we must remain ready to investigate further.

Figure 54. The male fiddler crab [113] uses its highly colored single giant claw for signaling, threats, and attracting the opposite sex.

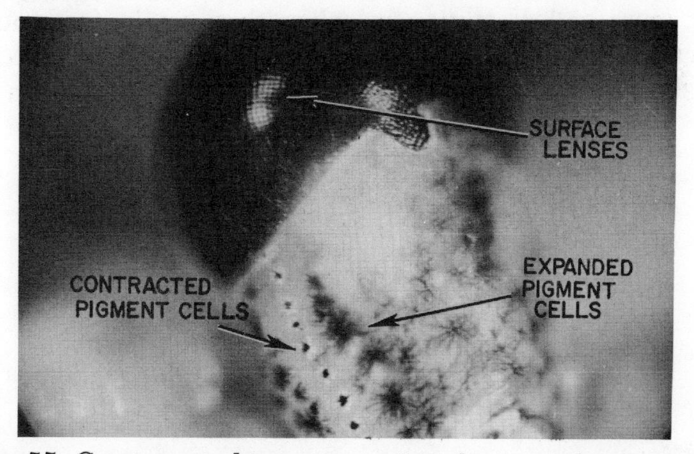

Figure 55. Crustaceans have very active chromatophores, and an easy place to see these function is on the stalk of a compound eye. This is the eyestalk of a shrimp.[88]

Many marine and aquatic invertebrates also use visual signals to a considerable extent—for example, the bright coloring of claws assumed by many crabs as a warning or in some cases attraction, and the fiddler crab's [113] waving of its giant claw to impress a female or an enemy. Many deep-water crustaceans also have photophores that they flash in the same way as fish, but apart from identifying their species and helping them to swarm, the meanings of the signals are not understood. More is known about crustaceans' color change for camouflage and courtship.

Fish living in the dark depths and using luminous organs have devoted their eye development to obtaining the greatest possible sensitivity and size, and this trend has also been followed by crustaceans. This is seen in deep-sea shrimps.[1, 101] Some luminous shrimps have developed quite a trick for deceiving fish that prey on them. They flash their line of light organs one after another along the line, making the shrimps appear to be moving at a different speed from the speed they are really going or making them appear to be moving when they are stationary.

Flashing their lights in another sequence—just single lights in series from the front to the rear—they might appear more or less stationary when they are in fact moving forward. If the lights are flashed in the opposite direction, the shrimps will appear to be moving much faster than they are.

13. Insect Antennae and the Use of Odors

Insect Antennae. Sometimes called "feelers," antennae are remarkable organs. They touch and feel objects, tap messages on other members of a group, pick up messages over great distances, act as direction finders, detect odors when only a few molecules are present in the air, and register temperature, humidity, and the carbon dioxide content of the air. No other organ in the animal world can do so many things, although the lateral-line system of fish must come close.

In most insects an antenna consists of three parts: a large base part attached to the head, called a *scape;* a short part, called a *pedicel,* that contains a sensitive organ known as Johnston's organ, which measures air currents and makes flight adjustments for wind; and a long, segmented part called a *flagellum* or *clavola.* These antennae are moved by muscles in the scape and pedicel, and some moths and gnats have featherlike plumes on their antennae (see Figure 57).

The antennae of grasshoppers and bees carry organs of smell that are often just tiny pits. These pits vary greatly in number. The queen bee has up to 2,400 of them, workers have about 1,600, and the drones up to 37,800. Some of these pits react to many different odors, others only to certain special odors. This is also true in moths, the males responding especially to female pheromones, which they can track over a distance of several miles when a favorable breeze carries the odor to them.

Certain other pits in bees' antennae are possibly hearing organs, but just how well these function is not known. Hairs on the antennae are sensitive to touch. There is a special name for the sense organs of insects—*sensillas*—and those for odor are exceedingly sensitive but only to odor molecules. There are four main kinds of sensillas on antennae, many of them for the detection of food or prey but perhaps most for picking up pheromones. Each kind of sensilla seems to differ in form in each kind of insect, and

Figure 56. Grasshoppers and locusts have antennae of many segments, with countless organs of smell.

as each sensillum sends a nerve fiber to the brain, an insect's entire behavior pattern is probably controlled by the number of sensillas on an antenna.

Some antennae are short and carry only a few sensillas; this is true in some larvae. Antennae can, on the other hand, be long and threadlike, as in the cockroach; shorter but densely packed with sensillas, as in the bee; or leaf-shaped with a feathery structure, as in many moths. The male

Figure 57. Some moths have featherlike antennae which may contain up to 150,000 sense cells.

polyphemus moth[106] has more than 60,000 sensillas on the branched part of its gigantic antenna, and these have 150,000 receptor cells, 60 to 70 percent of which are sensitive only to the female sex pheromone and 20 percent sensitive to other odors. The female antennae do not have sensillas for sex odor, but they have many more for general odor detection.

The Use of Odors by Insects. A great many insects use odor more than anything else for keeping in touch with each other. But we still have little idea of just how complicated these signals are or even how sensitive many of the receiving sensillas are. Many kinds of odors are used by the countless insect species, and we can only consider a very few of them.

PHEROMONES. These are probably the most important and plentiful of the odors that guide insect life. There appear to be nine principle ways in which insects respond to pheromones: alarm, species attraction, sexual attraction, swarming, grooming, exchanging solid food, exchanging liquid, recognition of others from the same nest, and recognition of caste or rank within the group. Only insects that live together in large numbers, such as ants and bees, have

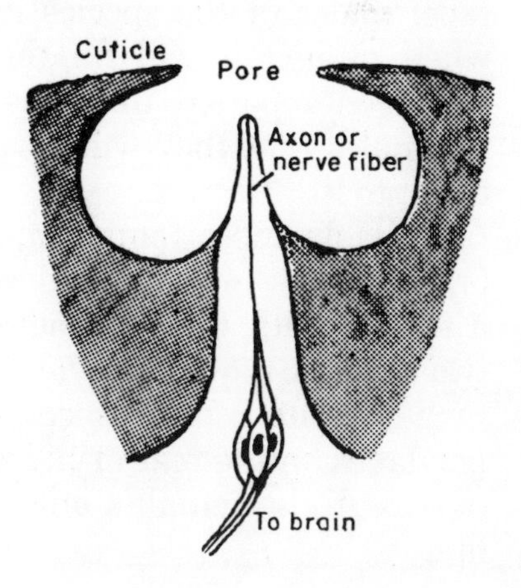

Figure 58. A *sensilla* odor receptor drawn from the locust.[63]

so many pheromones. Nonsocial insects concentrate mostly on sex attraction and alarm signals.

It has been said that alarm signals outnumber all others, but the sex pheromones must run very close to them in numbers. The most important thing about all of them is their concentration, because they all take time to fade. A pheromone's strength is therefore related to its age, and this will affect the behavior of any insect receiving it. All pheromones are unique to the species using them. Often the pheromones are so refined that they are unique only to a particular colony of the species, like the speech of people in two separated villages—one in Vermont and another in Tennessee.

A colony odor develops quite naturally in some insects; and ants, which are always exchanging body fluids with each other, are a good example of this. While considering ants another interesting phenomenon can be mentioned. Some ants release an alarm message by odor that acts to alarm other ants within the immediate vicinity but attracts those picking it up at a greater distance. This brings them to the danger spot to help the others deal with the emergency. This then is a double-purpose signal.

We see a similar signal in the sugar-beet wireworm.[62] The sex pheromone produced by its adult virgin flies will repel males of the species if it is in pure extract form, but when diluted it will fill them with intense excitement. It has been reported that this pheromone is so attractive to the males that they will cover twelve meters in ten seconds to get to it.

Probably most female moths give off a pheromone that brings males to them; indeed, many male moths can smell a female with their feathered antennae from a distance of two miles. Unique behavior can be seen in the queen butterfly[39] and its related species: the male uses brushlike appendages on the rear of his abdomen to deposit a pheromone dust onto the female's antennae, inducing her to mate with him.

The sensillas on the antennae of locusts must be exceedingly sensitive because locusts seem to be able to smell food from great distances. This could be why they are able to reach and destroy so much vegetation. Perhaps this is also possible to many other species of insects, but it seems that most species are more sensitive to their species pheromone than to the odor of food.

There are many insects in which both males and females transmit their own pheromones, which can be identified only by the opposite sex. This is well known in katydids and fireflies. One of the mysteries of pheromones is that they only attract—and are probably only recognized by—the species that uses them. If they have any effect on another species, it seems that it may be to repel them rather than attract them. This is just as well, because if pheromones were attractive to all insects, they would enable one species to prey on another too easily.

THE BEES' ODORS. The honeybee has other ways of using odor. It can carry it back to the hive both internally and on its body. The antennae of other hive members can get the odor of nectar clinging to the bee, and it can distribute samples of nectar from its stomach. At the same time the antennae of other bees can check the hive odor on the bee's body, check its individual odor secreted by its abdominal scent glands, and sense a track-marking odor from its mandibular glands. This is a great deal of information to be obtained at one time from one insect. When we remember the bee's dances and the sounds it makes, we can see that with so many ways of communicating, there is every reason for this insect to have such a highly organized way of life.

What has been said in the foregoing chapters is but a small fraction of the story of animal communication, much of it gathered from the bibliography that follows. For the student who wishes to have a more thorough knowledge of communication in animals and wishes to study the many technical details that are available, R. G. Busnel's book, *Acoustic Behavior of Animals,* is an excellent starting point.

Glossary of Scientific Names

1. Acanthepheridae
2. *Acherontia atropos*
3. *Acridotheres tristis*
4. *Aepyceros melampus*
5. *Afrixulus* spp.
6. *Agkistrodon mokasen*
7. *Alouatta palliata*
8. Alpheidae
9. *Alytes obstetricans*
10. *Amblyopsis spelaeus*
11. *Amphibolurus barbatus*
12. *Aneides lugubris*
13. *Anoptichthys jordani*
14. *Antilope cervicapra*
15. *Aotus trivirgatus*
16. Arctiidae
17. *Arthrosphaera aurocineta*
18. *Ascaphus*

19. *Balaenoptera physalus*
20. *Balistes conspicillum*
21. *Balistes* spp.
22. *Balistes vetula*
23. *Bitis arietans*

24. *Callicebus moloch*
25. *Callorhynchus callorhynchus*
26. Cantharides
27. Carangidae
28. *Chiroderma* spp.
29. *Chlamydosaurus kingi*
30. *Clarias batrachus*
31. Clupeoidea
32. *Colobus guereza*
33. *Crotophága* spp.
34. *Cygnus cygnus*
35. *Cynomys ludovicianus*
36. *Cynoscion* spp.
37. Cypriniformes
38. Cyprinodonts

39. *Danaus gilippus*

40. *Electrophorus electricus*
41. Engraulidae
42. *Eubalaena glacialis*

43. *Fregata aquila*

44. *Gadus aeglefinus*
45. Geometridae
46. *Grallina* spp.
47. *Gymnarchus*

48. *Haemulon* spp.
49. *Hemidactylus frenatus dum*
50. *Heteropneustes fossilis*
51. *Hipposideros diadema*
52. Holocentridae
53. *Holocentrus ascensionis*
54. *Hyperolius* spp.
55. *Hypsignathus monstrosus*

56. *Kakatoe galerita*

57. *Lachesis muta*
58. Lampyridae
59. *Lampyris noctiluca*
60. *Lemur catta*
61. *Lialis burtoni*

62. *Limonius californicus*
63. *Locusta migratoris*
64. *Lucia lediacolis*
65. *Lucioperca lucioperca*

66. *Magicicada cassini*
67. *Magicicada cassius*
68. *Magicicada septemdecim*
69. *Marmota flaviventris*
70. *Megaptera novaeangliae*
71. *Megalobatrachus* spp.
72. *Melanogrammus aeglefinus*
73. *Melichthys* spp.
74. *Melospiza melodia*
75. *Menura superba*
76. *Meriones unguiculatus*
77. *Micropogon undulatus*
78. *Micrurus* spp.
79. *Mirounga leonina*
80. Mormyridae
81. *Mycteroperca bonaci*
82. *Mysticete* spp.

83. Noctuidae
84. *Nycticorax nycticorax*

85. *Opsanus tau*
86. *Orcinus orca*
87. *Otis tarda*

88. *Palaemonetes vulgaris*
89. Palinurids
90. *Papio* spp.
91. *Parus* spp.
92. *Phalacrocorax carbo*
93. *Photinus* spp.
94. *Phyllurus platurus*
95. *Platycerus eximius*
96. *Pogonias chromis*
97. *Porychthys notatus*
98. *Pungitius pungitius*

99. *Rana cavitympanus*

100. Sciaenidae
101. Sergestidae
102. *Solenodon paradoxus*
103. *Steatornis caripensis*
104. *Stygicola dentatus*

105. *Talpa* spp.
106. *Telea polyphemus*
107. *Testudo elephantopus*
108. *Tettigonia viridissima*
109. *Thamnophis* spp.
110. *Thryothorus endovicianus*
111. *Prionotus*
112. *Tyto alba*

113. *Uca crenulata*
114. *Uroaëtus audax*
115. Urodela

116. *Zalophus californianus*

Bibliography

Agrawal, V. P. and Sharma, R. S., "Sound Producing Organs of Indian Catfish (*Heteropneustes fossilis*)." Ann. Mag. Nat. Hist. 8(89/90), 1965.

Allen, Glover M., *Birds and Their Attributes*. New York, Dover, 1962.

Armstrong, Edward A., *Bird Display and Behavior*. New York, Dover, 1965.

——— *The Way Birds Live*. New York, Dover, 1967.

Beebe, C. William, *The Bird*. New York, Dover, 1965.

Bernhard, Carl G., *The Functional Organization of the Compound Eye*. Oxford, Pergamon Press, 1966.

Bowers, J. M., and Alexander, Bruce K., "Mice; Individual Recognition By Olfactory Cues." *Science*, Vol. 158 (December 1, 1967), pp. 1208–10.

Brown, J. L., "Vocalization Evoked from the Optic Lobe of a Songbird." *Science*, Vol. 149 (August 27, 1965), pp. 1002–3.

Buck, J. and E., "Mechanism of Rhythmic Synchronous Flashing of Fireflies." *Science*, Vol. 159 (March 22, 1968), pp. 1319–27.

Bunak, V. V., and Yakimenko, I. A., *Imitative and Pantomimic Reactions of Cercopithecan Monkeys and Chimpanzees*. Moscow, Nauka, 1965.

Burkhardt D., and others, *Signals in the Animal World*. New York, McGraw Hill, 1968.

Busnel, R. G., ed., *Acoustic Behavior of Animals*. London and New York, Elsevier Publishing Company, 1963.

Cogger, Harold, *Australian Reptiles*. Sydney, A. H. & A. W. Reed, 1967.

Conley, Robert A. M., "Locusts: Teeth of the Wind." *National Geographic Magazine*, Vol. 136 (August 1969), pp. 202–227.

Curtis, Brian, *The Life Story of the Fish*. New York, Dover, 1961.

Duellman, William E., "Social Organization in the Mating Calls of Some Neotropical Anurans." Amer. Midland Natur., Vol. 77 (1967).

Eisenberg, John F., "The Behavior of *Solenodon Paradoxus* in Captivity." Zoologica, New York, 51 (1), 1966.

Ekman, P., and others, "Pan-cultural Elements in Facial Displays of Emotion." *Science*, Vol. 164 (April 4, 1969), pp. 86–8.

Evans, William F., *Communication in the Animal World*. New York, Crowell, 1968.

Fossey, Dian, and Campbell, Robert M., "Making Friends with Mountain Gorillas." *National Geographic*, Vol. 137 (January 1970), pp. 48–67.

——— "More Years with Mountain Gorillas." *National Geographic*, Vol. 140 (October 1971), pp. 574–84.

Gilbert, Bill, *How Animals Communicate*. New York, Pantheon Books, 1966.

Gould, J. E., and others, "Communication of Direction by the Honey Bee." *Science*, Vol. 169 (August 7, 1970), pp. 544–54.

Grimes, L. G., "Antiphonal Singing . . ." *Science,* Vol. 108 (1966).

Hanson, F. E., and others, "Synchrony and Flash Entertainment of the New Guinea Firefly." *Science,* Vol. 174 (October 8, 1971), pp. 161–4.

Harrison, J. M., and Irving, R., "Visual and Non-Visual Auditory Systems in Mammals." *Science,* Vol. 154 (November 11, 1966), pp. 738–43.

Hawkins, A. D., and Chapman, C. J., *Underwater Sounds of the Haddock (Melanogrammus aeglefinus).* J. Marine Biol. Ass., United Kingdom, 46(2), 1966.

Hinde, L. A., *Monkey Communication.* London, Phil. Trans. Roy. Society, 251/722.

Howes, P. E., "Mechanism of the Insect Ear." *Nature,* Vol. 213 (January 1967), pp. 367–9.

Hyman, Libby H., *Comparative Vertebrate Anatomy.* Chicago, University of Chicago Press, 1956.

Idyll, C P., *Abyss.* New York, Crowell, 1964.

Immelman, Klaus, "Behavioral Studies of the Kangaroos in Australia." Zool. Gart., 31 (3/4), 1965.

Jacobson, M., and others, "Sex Attractant of the Sugar Beet Wireworm." *Science,* Vol. 159 (January 12, 1968), pp. 208–10.

Johnson, D. L., "Honeybees—Do They Use Distance/Direction Information Contained in their Dance Maneuver?" *Science,* Vol. 155 (February 1967), pp. 844–7.

Johnson, M. E., and Snook, H. J., *Seashore Animals of the Pacific Coast.* New York, Dover, 1967.

Jolly, A., "Lemur Social Behavior and Primate Intelligence." *Science,* Vol. 153 (July 29, 1966), pp. 501–6.

Kellogg, W. N., "Communication and Language in the Home Trained Chimpanzee." *Science,* Vol. 162 (October 25, 1968), pp. 423–7.

Larousse, *Encyclopedia of Animal Life.* London and New York, Hamlyn, 1971.

Lawick-Goodall, Jane van, "My Life Among Wild Chimpanzees." *National Geographic,* Vol. 124 (August 1963), pp. 272–308.

———, "New Discoveries Among Africa's Chimpanzees." *National Geographic,* Vol. 128 (December 1965), pp. 802–831.

Lissmann, H. W., "Electric Location by Fishes." *Scientific American,* Vol. 208 (March 1963), pp. 50–9.

Lloyd, James E., *Studies on the Flash Communication System in Photinus Fireflies.* Pub. of Mus. of Zoology, University of Michigan, 1966.

Maier, N. R. F., and Schneirla, T. C., *Principles of Animal Psychology.* New York, Dover, 1964.

Marler, P., and Hamilton, W. J., *Mechanisms of Animal Behavior.* New York, Holt, Rinehart & Winston, 1965.

Marler, P., *Primate Behavior: Field Studies of Monkeys and Apes.* I. De Vore, ed., New York, Holt, Rinehart & Winston, 1965.

———, "Animal Communication Signals." *Science,* Vol. 157 (August 18, 1967), pp. 769–74.

Marshall, N. B., *The Life of Fishes.* London, Weidenfeld & Nicholson, 1965.

McGill, T. E., ed., *Readings in Animal Behavior.* New York, Holt, Rinehart & Winston, 1965.

Morris, Desmond, *The Mammals.* London, Hodder & Stoughton, 1965.

Moynihan, M., *The Night Monkey (Aotus trivirgatus).* Smithsonian Miscellaneous Collection, 146(5), 1964.

Mulligan, J. A., *Singing Behavior and Its Development in the Song Sparrow.* University of California Publications in Zoology, Vol. 81, 1966.

National Geographic Society, *Wondrous World of Fishes.* Washington, D.C., 1965.

———, *The Marvels of Animal Behavior.* Washington, D.C., 1972.

Noble, G. Kingsley, *The Biology of the Amphibia.* New York, Dover, 1955.

Ommanney, F. D., *A Draught of Fishes.* London, Longmans, 1965.

Payne, Roger S., and McVay, S., "Songs of Humpback Whales." *Science,* Vol. 173 (August 13, 1971), pp. 585–97.

Plisk, T. E., and Eisner, T., "Sex Pheromones of the Queen Butterfly." *Science,* Vol. 164 (June 6, 1969), pp. 1170–2.

Prince, J. H., *Comparative Anatomy of the Eye.* Springfield, Illinois, Charles C. Thomas, 1956.

———, *Animals in the Night.* New York, Thomas Nelson, Inc., 1971.

———, *The Universal Urge.* New York, Thomas Nelson, Inc., 1972.

Protasov, V. R., *Sound Signals in Fish.* Moscow, Bionika Nauka, 1965. & Nauk. SSSR Ser. Biol. 31(1), 1966.

Reid, K. H., "Periodical Cicada: Mechanism of Sound Production." *Science,* Vol. 172 (May 28, 1971), pp. 949–51.

Roeder, Kenneth D., "Moths and Ultrasound." *Scientific American,* Vol. 212 (April 1965), pp. 94–102.

Scott, John Paul, *Animal Behavior.* Chicago, University of Chicago Press, 1972.

Schneider, D., "Insect Olfaction." *Science,* Vol. 163 (March 7, 1969), pp. 1031–7.

Schusterman, R. J., and others, "Underwater Vocalization by Sea Lions." *Science,* Vol. 154, (October 28, 1966), pp. 540–2.

Schusterman, R. J., and Dawson, R. G., "Barking Dominance and Territoriality in Male Sea Lions." *Science,* Vol. 160 (April 26, 1968), pp. 434–7.

Simmons, J. A., and others, "Periodical Cicada: Sound Production and Hearing." *Science,* Vol. 171 (January 15, 1971), pp. 212–3.

Smith, W. J., "Messages of Vertebrate Communication." *Science,* Vol. 165 (July 11, 1969), pp. 145–50.

Smythe, R. H., *Animal Habits.* Springfield, Illinois, Charles C. Thomas, 1962.

Snodgrass, R. E., *Insects.* New York, Dover, 1967.

———, *Principles of Insect Morphology.* New York, McGraw Hill, 1935.

Storer, Tracy I., *General Zoology.* New York, McGraw Hill, 1943.

Swenson, Melvyn J., ed., *Duke's Physiology of Domestic Animals.* Ithaca, New York, Comstock Publishing Association, 1970.

Thorpe, W. H., and Hinde, R. A., *Bird Vocalizations.* New York, Cambridge University Press, 1969.

Todd, J. M., and others, "Chemical Communication in Social Behavior of a Fish." *Science,* Vol. 158 (November 3, 1967).

Van Tets, G. F., Aust. Nat. Hist., 15/1, (March 1965).

Vilks, E. K., *Results of Experimental Studies of Complex Behavior Patterns of Birds in Natural Conditions.* Moscow, Nauka, 1965.

Von Frisch, K., and others, "Honeybees: Do They Use Direction and Distance Information Provided by Their Dances?" *Science,* Vol. 158 (November 24, 1967), pp. 1072–7.

Washburn, S. L., and others, "Field Studies on Old World Monkeys and Apes." *Science,* Vol. 150 (December 17, 1965), pp. 1541–7.

Waterman, T. H., *The Physiology of Crustacea,* Vol. 2. New York, Academic Press, 1961.

Wenner, A. M., and others, "Honey Bee Recruitment to Food Sources." *Science,* Vol. 164 (April 4, 1969), pp. 84–6.

———, "Sound Communication in Honeybees." *Scientific American,* Vol. 210 (April 1964), pp. 116–22.

Whitten, K., and others, "Estrus-inducing Pheromone of Male

Mice." *Science,* Vol. 161 (August 9, 1968), pp. 584–5.

Wilson, E. O., "Chemical Communication in the Social Insects." *Science,* Vol. 149 (September 3, 1965), pp. 1064–71.

Zoological Society of London, *Evolutionary Aspects of Animal Communication.* Symposium, 1962.

Index